MATERIALS SCIENCE AND TECHNOLOGIES

CARBON NANOTUBES

SELECT ARMY RESEARCH LABORATORY STUDIES

Library of Congress Cataloging-in-Publication Data

ISBN: 978-1-62618-493-0

Published by Nova Science Publishers, Inc. † New York

CONTENTS

PREFACE

This book provides an overview of select army research laboratory studics on carbon nanotubes. Topics include increasing the capacitance of carbon nanotube (CNT)-or graphene-based supercapacitors by adding pseudocapacitive manganese oxide nanoparticles; electrochemical double layer capacitors fabricated using carbon nanotube (CNT)/paper flexible electrodes; improving microbolometric response using carbon nanotubes; nanomagnetics-magnetic nanoparticles filled carbon nanotubes; and the performance of carbon nanotubes in extreme conditions and in the presence of microwaves.

Chapter 1 - Supercapacitors are of interest because they are energy storage devices with greater power density than batteries. Unfortunately, supercapacitors have significantly lower energy densities than batteries, which limit the applications for supercapacitors. The addition of pseudocapacitance, that is, chemical reactions similar to those in batteries but which behave electrically like a capacitance, can be used to increase the energy density of supercapacitors. This work focuses on increasing the capacitance of carbon nanotube (CNT)- or graphene-based supercapacitors by adding pseudocapacitive manganese oxide nanoparticles. A number of methods have been investigated for fabrication of CNT/graphene/manganese oxide composites. Manganese oxide nanoparticle pseudocapacitance has been successfully incorporated into CNT/graphene-based supercapacitors. Further optimization of the composite compositions and electrode fabrication methods is still required to optimize the energy densities of these devices.

Chapter 2 - Electrochemical double layer capacitors are fabricated using carbon nanotube (CNT)/paper flexible electrodes. An extensive electrode fabrication study was conducted, resulting in single electrode specific with GO is that it is water soluble and once the electrode has been fabricated, the GO can be thermally reduced to graphene (G, or rGO). Typically, 225 °C overnight (in air) has been used to thermally reduce the GO used here. The thermal conversion of manganese acetate (MnAc) (obtained from Sigma Aldrich) has been used to produce the MnOx NPs used here [3]. Higher temperatures (300 °C/1 h/in air) have typically been used to convert the MnAc to MnOx. Thermal gravimetric analysis (TGA) was used to determine that the MnAc decomposes below 300 °C, as shown in figure 1. The amount of mass lost indicates that Mn_3O_4 may be the resulting species. The electrode materials produced were examined with scanning electron microscopy (SEM).

2.2. G/MnOx NP Synthesis Method Investigation

2.2.1. Dry Mixing/Grinding

Dry mixing and grinding of CNTs with MnAc was attempted as described by Yi Lin [4]. After baking the dry mixture, SEM imaging indicated there was insufficient mixing of the materials and that using more than 5 mole % of MnAc may be useful, as can be seen in figure 2.

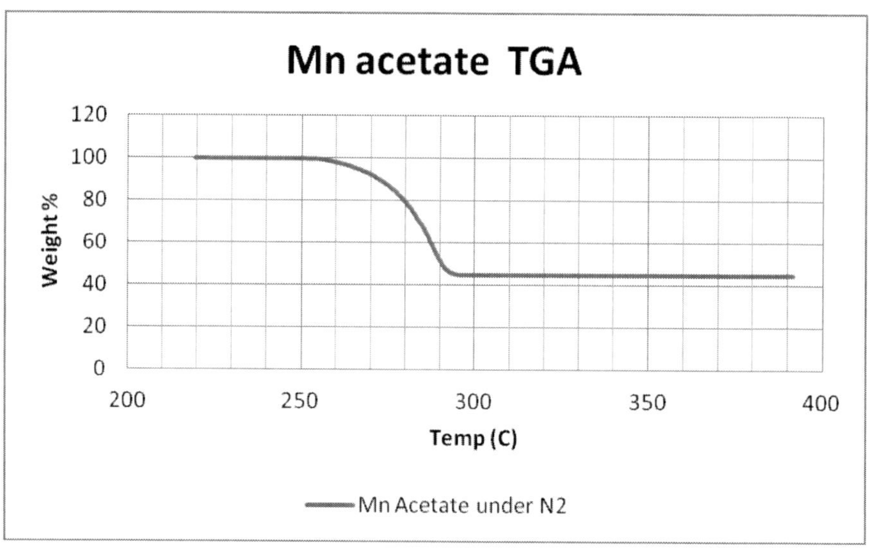

Figure 1. TGA at 1°C/min of MnAc demonstrating that it decomposes below 300 °C. The mass lost indicates that Mn3O4 may be the resulting species.

Figure 2. SEM image of ball milled MnAc (3 mole %) with multi-wall carbon nanotubes (MWCNTs) annealed at 370 °C. The resulting material is rather inhomogeneous and the MnOx NPs do not appear to be well associated with the MWCNTs (Mn370C303).

2.2.2. Solution Mixing Synthesis

In order to obtain a more thorough mixing of the CNTs and MnAc, a mixing of these materials in solution was attempted. A mass ratio of 18:82 MnAc:MWCNT was suspended in ethanol (the MnAc dissolves, but the MWCNTs are merely suspended).

The resulting suspension was immediately drop cast after ultrasonication and the resulting deposit was baked at 370 °C. As can be seen in figure 3, 10-nm NPs are associated with the MWCNTs, but there are also agglomerated MnOx NPs. The lack of complete mixing of the materials is likely due to separation during the dropcasting.

2.2.3. Freeze-dry Synthesis

In an effort to prevent the segregation of the MnAc from the carbon during evaporation of the solution, a freeze dry method was used. In addition, a water solution of GO was used for the carbon source since the GO is actually in solution and not merely suspended as the MWCNTs above were.

Next, 2 mg of GO and 3.3 mg of MnAc were dissolved in 3 ml of water. This solution was flash frozen by depositing it drop wise into liquid nitrogen. The resulting ice crystals were freeze dried at 10^{-3} torr. After baking to reduce the GO to rGO and convert the MnAc to MnOx, the resulting powder was imaged with the SEM.

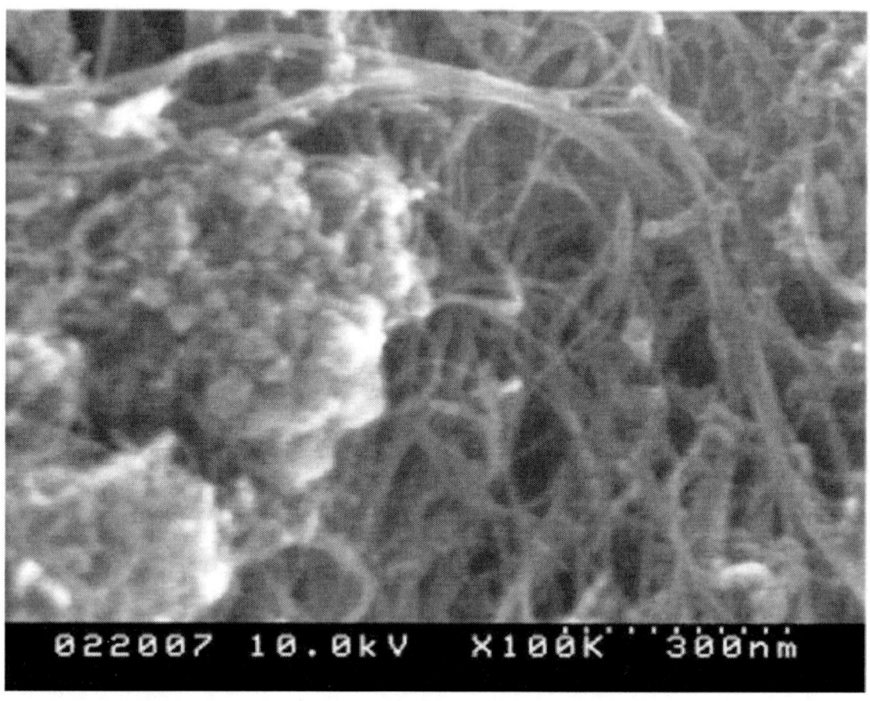

Figure 3. SEM image of ball milled MnAc (3 mole %) with MWCNTs annealed at 370°C. The resulting material is rather inhomogeneous and the MnOx NP do not appear to be well associated with the MWCNTs (Mnsoln370107).

Figure 4 shows there are small MnOx NPs coating the rGO. While successful in producing NP-coated graphene, it was difficult to collect sufficient material for further electrode processing, so additional synthesis methods were attempted.

2.2.4. Hydrothermal Processing

Next, 2 to 1 solutions (by mass) of MnAc and CNTs or GO were hydrothermally treated at 200 °C for 14 h and the resulting MnOx NP decorated materials are seen in figure 5.

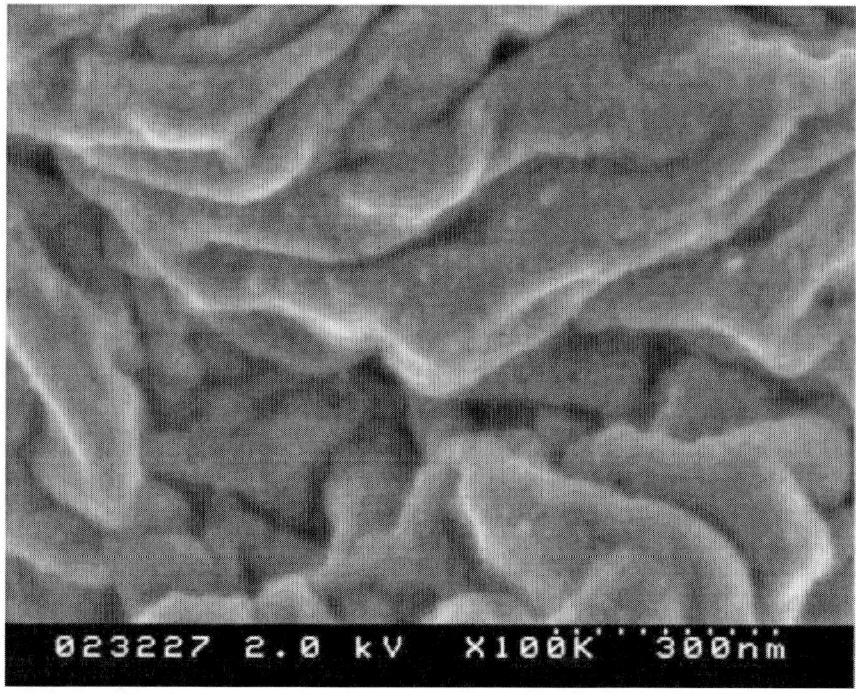

Figure 4. SEM of rGO coated with MnOx nanoparticles fabricated through freeze-dry method (VNRA1).

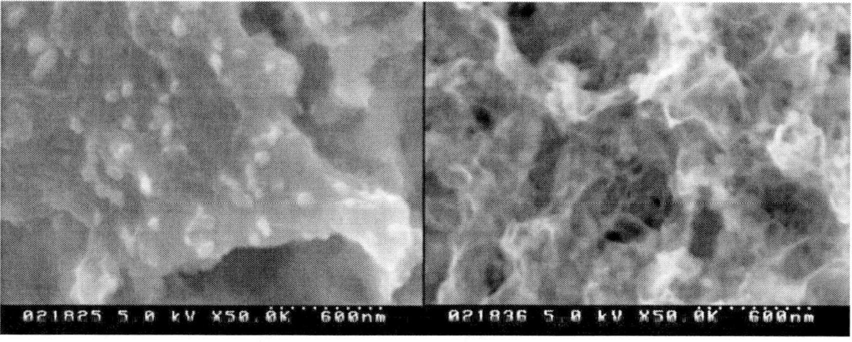

Figure 5. Hydrothermally produced MnOx NPs on CNTs (left) and rGO (right) (MnHTCNT204, MnHTG205).

For subsequent hydrothermal syntheses, a MnAc and GO solution was mixed with trace amounts of potassium hydroxide (KOH) using a procedure described by Wang [5].

The fabrication process involved stirring 0.4 mmol MnAc solution, 0.9 mmol KOH solution, and GO solution (1 mg/mL) and heating at 180 °C for 14 h in a Teflon-coated autoclave. The solution was then deposited onto current collectors by dropcasting. These electrodes were then baked at 300 °C for 1 h to ensure that the MnAc was converted to MnOx NPs.

2.2.5. Spray Dry Processing

The final synthesis method investigated was spray drying, which was done with the assistance of Prof. Hongwei Qiu of Stevens Institute of Technology. As it turns out, spray drying has some interesting effects on the GO all by itself even before MnAc is incorporated.

In spray drying, the GO solution is aerosolized and the droplets are dried with a hot carrier gas. The surface tension of the drying droplets, crumples the GO sheets up into nanospheres. Figure 6 shows the flat uncrumpled appearance of drop cast GO solution and the crumpled morphology of the spray dried GO solution.

The crumpled morphology is of interest for further investigation since it may result in more surface area and greater ionic conductivity (porosity) in the electrodes produced this method.

On the other hand, the crumpled morphology will increase the electrical resistivity due to reduced contact between adjacent rGO sheets in the final electrode.

When MnAc is included with GO in the spray-dried solution, a good mixture of the MnOx NP and rGO sheets results, as can be seen in figure 7. The figure shows 5- to 10-nm NPs on the rGO.

Figure 6. SEM images of reduced, drop-cast GO solution (left) and reduced, spray-dried GO solution (right) (MHE23502 and MHE23802).

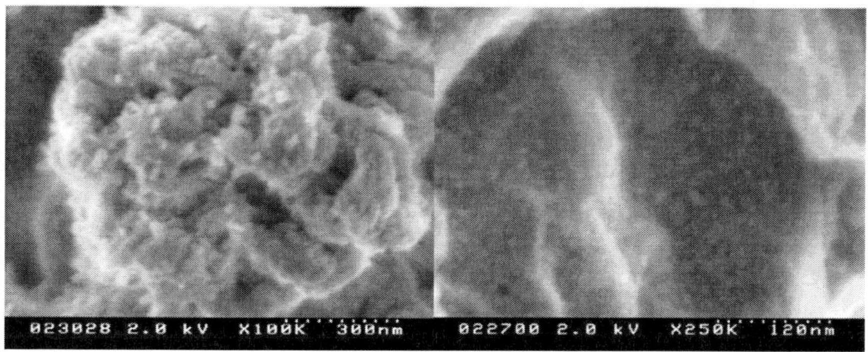

Figure 7. SEM of MnOx NPs formed on rGO using spray dry process (SDGO14, SDGO09).

2.3. Capacitor Performance

2.3.1. Capacitor Evaluation

Electrodes were fabricated by drop casting solutions containing the GO/CNT/MnAc materials onto titanium (Ti) or stainless steel current collectors and then annealing them to reduce the GO or convert the MnAc to MnOx as needed. Testing was performed using cyclic voltammetry (CV) and electrochemical impedance spectroscopy (EIS) on a potentiostat typically using a three-electrode setup with a large excess of electrolyte. In a later experiment, a two electrode setup was used with fabricated button cells. The CV scan rate for the majority of the testing was 20 mV/s, except where noted. The voltage range of scan was 0 to 0.7 V (versus the silver/silver chloride electrode when using the three electrode mode). The electrolyte typically used in this study was 0.5 M potassium sulfate (K_2SO_4). A typical CV obtained is seen in figure 8.

The specific capacitance is calculated from a three electrode measurement using

$$\text{CV Current}/(\text{Scan Rate}*\text{Mass of Sample}) = \text{Capacitance (F/g)} \quad (1)$$

where the current is measured as half the difference between the oxidation and reduction currents near the center of the CV curve near where this value is a minimum to avoid overstating the capacitance. In a two electrode (button cell) measurement, the specific capacitance is calculated in the same manner, but the mass of both electrodes is included and the result is multiplied by four to report it in the three electrode standard as is observed in the literature.

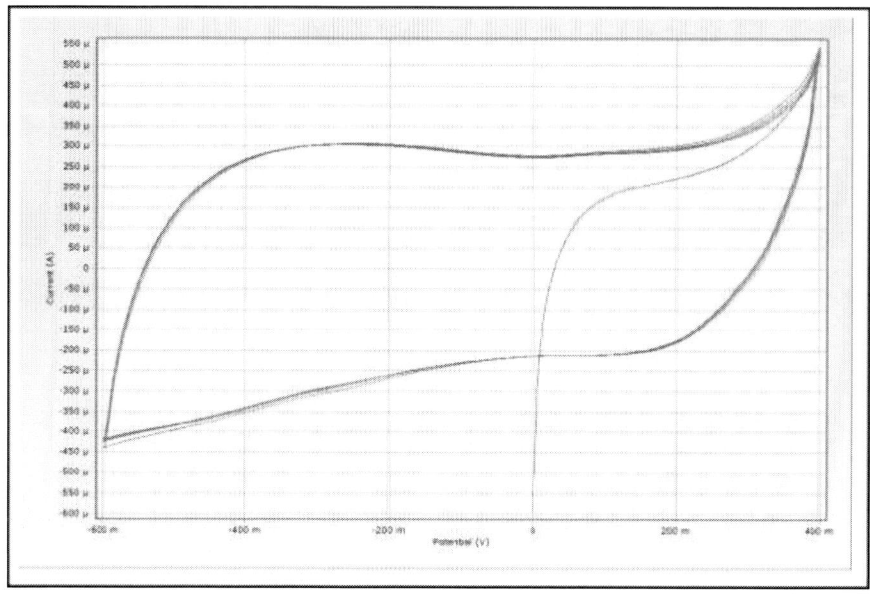

Figure 8. Representative CV scan of a rGO/MnOx NP electrode in 0.5 M K_2SO_4.

2.3.2. Capacitance Increasing with Cycling

A MnOx NP decorated MWCNT material was made using a spray-dry process. This electrode material was deposited onto a current collector from an n-methylpyrrolidone (NMP) suspension that also contained polyvinylidene fluoride (PVDF) binder. When this electrode was cycled for 1200 cycles, the capacitance was observed to increase from 29.2 to 42.8 mF, as shown in figure 9. After subtracting out the capacitance expected from the MWCNTs alone, the specific capacitance of the MnOx NPs was calculated to increase from 8.3 to 31.8 F/g. It is believed that this increase is due to the conversion of the MnOx from its initially synthesized phase believed to be Mn3O4 to the more electrochemically active birnessite phase [6].

2.3.3. Morphology Change with Cycling

Unfortunately, the capacitance does not always increase with cycling. Often, it decreases for various reasons. There may be a potential issue with the MnOx dissolving during part of the CV cycle and then redepositing later in the cycle. Figure 10 shows the morphology of the MnOx NP after 9000 cycles.

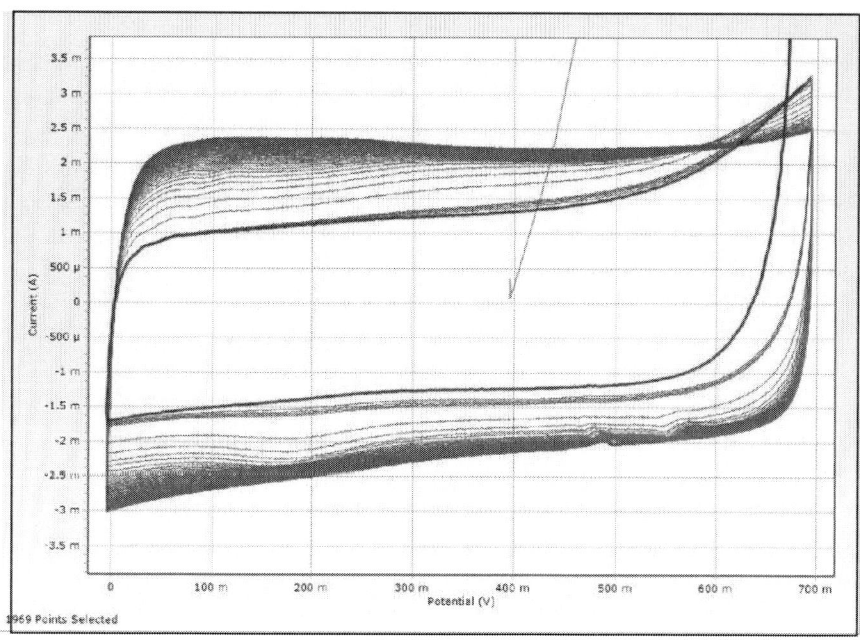

Figure 9. Depiction of 1200 CV cycles of a spray-dried MWCNT/MnOx NP electrode made with Kynar 761 binder. The red trace is the first cycle (MHE23403).

In this case, the capacitance was still increasing, but the significant changes in morphology could be an issue in a practical device.

Since the MnOx dissolves during part of the CV cycle, it is possible that the electrodes tested in a three electrode mode using a large excess of electrolyte could lose a significant amount of Mn to the electrolyte. However, making a button cell with a limited amount of electrolyte was not successful in increasing the measured capacitance, so this may not be a serious issue with the measurements made here.

2.3.4. Mechanically Unstable Electrodes

The hydrothermal method used mixed KOH with MnAc and GO. The KOH seemed to have formed crystals on some of the dropcast electrodes. Here, KOH crystals were formed in the electrode and created an unstable electrode mass by absorbing moisture from the air. Dissolution of the KOH crystals also resulted in the electrode deteriorating in the electrolyte, producing falling capacitance readings.

Another potential reason for an electrode material losing electrical contact with the current collector is due to insufficient binder. In much of this work, a binder material was not used to avoid any complications that would result if it coated the surface of the electrode materials. However, electrochemical deterioration was observed when testing many electrodes, sometimes due to a detachment of the electrode material from the current collector. When necessary electrodes were made using a PVDF binder, but the formulation has not been optimized. The CNT and G materials have such large surface areas that more binder than is typically used to make electrodes is likely required to obtain adequate mechanical robustness.

Figure 10. SEM image of MWCNT/MnOx NP electrode after 9000 CV cycles showing MnOx NP platelet formation. (MHE20001).

2.3.5. Greater Capacitance at Slower Scan Rate

Since redox reaction based pseudocapacitance is slower than the double-layer capacitance contribution, it is worthwhile to measure how the capacitance changes with the scan rate. The electrode used was made from spray-dried powder from a solution that had 100 mg of GO and 145 mg of MnAc in 50 ml of water, which should yield >10 mole % Mn:C. This spray-dried GO/Mn Ac powder was baked at 275 °C for 4 h. Then, 1.06 mg of the reduced/converted powder was mixed with 0.37 mg of SWCNTs in NMP. The SWCNTs were added to improve the electrical conductivity between the MnOx NP covered G nanospheres and bind the electrode materials together. Figure 11 shows a similar electrode made without the MnOx NPs.

Figure 11. SEM of spray-dried rGO nanosphere electrode with 20% SWCNT as binder (MHE24303).

The resulting suspension was dropcast onto a Ti current collector. The capacitance as a function of scan rate is plotted in figure 12. At the slowest scan rate measured, 2 mV/s, the capacitance was measured to be 135 F/g.

2.3.6. Electrolyte Effects

MnOx containing electrodes are typically used with neutral (pH) electrolytes. In order to investigate the effect of the electrolyte on the capacitance observed, the same spray-dried rGO/MnOx NP with SWCNT electrode as above was tested at a 2 mV/s scan rate in three different electrolytes: 0.5 M K_2SO_4, 1 M sodium chloride (NaCl), and 1 M calcium chloride ($CaCl_2$).

The qualitative CV behavior of the three electrolytes can be seen in figure 13. Before taking these CVs, the electrode was cycled in 0.5 M K_2SO_4 for 200 specific capacitances for MnOx/G or CNT composite electrodes in the literature.

In future work, the best ratio of MnOx:G/CNT should be determined. In much of the reported MnOx:G/CNT composite electrode work, the G/CNT is merely an additive that increases the electrical conductivity of the MnOx. The energy and power requirements of the ultimate application are liable to determine the optimum ratio of materials. In addition, additional synthesis routes for producing the most electrochemically active MnOx phase should be investigated. Finally, more electrode fabrication process optimization is required to prevent electrode deterioration that is sometimes seen.

REFERENCES

[1] Pandolfo, G.; Hollenkamp, A. F. *J. Power Sources* 157, 11–27 (2006).

[2] Ervin, M. H.; Mailly, B.; Palacios, T. Electrochemical Double Layer Capacitance of Metallic and Semiconducting SWCNTs and Single-Layer Graphene. *ECS Transactions* 2012, 41 (22), 153–160.

[3] Xu, F.; Kang, B.; Li, B.; Du, H. Recent Progress on Manganese Dioxide Based Supercapacitors. *Journal Material Research* 2010, 25 (8) 1421–1432.

[4] Lin, Yi; Watson, K. A.; Fallbach, M. J.; Ghose, S.; Smith, J. G. Jr; Delozier, D. M.; Cao, W.; Crooks, R. E.; Connell, J. W. Rapid, Solventless, Bulk Preparation of Metal Nanoparticledecorated Carbon Nanotubes. *ACS Nano* 2009, 3, 871–884.

[5] Wang, D.; Li, Y.; Wang, Q.; Wang, T. Facile Synthesis of Porous Mn3O4 NanocrystalGraphene Nanocomposites for Electrochemical Supercapacitors. *Journal Inorganic Chemistry* 2012, 628–635.

[6] Inoue, R.; Nakashima, Y.; Tomono, K.; Nakayama, M. Electrically Rearranged BirnessiteType MnO2 by Repetitive Potential Steps and Its Pseudocapacitive Properties. *Journal of the Electrochemical Society* 2012, 159 (4), A445–A451.

In: Carbon Nanotubes
Editor: Percy Szalkowski

ISBN: 978-1-62618-493-0

Chapter 2

CARBON NANOTUBE BASED FLEXIBLE SUPERCAPACITORS*

Christopher M. Anton and Matthew H. Ervin

LIST OF SYMBOLS, ABBREVIATIONS AND ACRONYMS

CNT	carbon nanotube
CVs	cyclic voltammograms
ESEM	environmental scanning electron microscope
MWCNT	multi-walled CNT
SDBS	sodium dodecylbenzenesulfonate
SWCNTs	single walled carbon nanotubes

ABSTRACT

Electrochemical double layer capacitors are fabricated using carbon nanotube (CNT)/paper flexible electrodes. An extensive electrode fabrication study was conducted, resulting in single electrode specific capacitances in excess of 100 F/g. Surfactants and dispersants in the CNT suspension prior to CNT deposition onto paper substrates was found to be detrimental to electrode performance. Following electrode fabrication,

* This is an edited, reformatted and augmented version of an Army Research Laboratory publication, ARL-TR-5522, dated April 2011.

three full cell packaged supercapacitor architectures were investigated. One architecture used a liquid KOH electrolyte, while the other two architectures used a PVA:KOH gel electrolyte. Devices were tested under static bending conditions, and proved to function under bending.

1. BACKGROUND

Electrochemical double layer capacitors (Supercapacitors) are expected to play a significant role in future hybrid power systems due to their high specific power, cycle life, and tolerance of extreme environmental conditions [1]. The development of flexible, conformable energy storage devices is also of great interest to the Army due to ease of packaging and coupling with flexible electronics. In supercapacitors, two electrodes are placed in an electrolyte and a voltage is applied between the two electrodes. During the application of voltage, an electrochemical double layer is formed on both electrodes when negative ions from the electrolyte are drawn to the positive electrode, and positive ions are drawn to the negative electrode. When these ions are released by opening a path between the positive and negative electrodes, the energy stored in the double layers is discharged through the external circuit.

Because supercapacitors store charge only on the electrode surfaces, maximizing the amount of surface area that is accessible to the electrolyte ions is of critical importance. Conventional supercapacitors typically use activated carbon electrodes on top of a current collector. Carbon is an excellent electrode material due to a variety of favorable chemical properties, including chemical stability and large electrochemical windows in a variety of electrolytes [1]. The advantage of activated carbon is its low cost relative to competing electrode materials. The disadvantage is its modest specific capacitance (capacitace per mass), due to the fact that electrolyte ions cannot collect between the individual graphene layers of the activated carbon electrode.

Carbon nanotube (CNT) electrodes have been thoroughly investigated as supercapacitor electrodes over the past decade. CNTs are an appealing electrode material due to their high conductivity, and the fact that single walled carbon nanotubes (SWCNTs) are comprised entirely of surface atoms that could interact with electrolyte ions. Another advantage of CNT electrodes is their ability to bend and flex. Activated carbon electrodes typically require binders and conduction additives that do not contribute to overall device capacitance and add weight to the system. CNT electrodes do not typically require any type of binder. Several different methodologies have immerged in the literature to fabricate flexible CNT based supercapacitor electrodes. CNTs have been spin coated [2, 3], filter deposited [4], spray deposited [5], or directly grown through chemical vapor deposition (*6–8*) onto a variety of flexible substrates including plastics and carbon cloth. More recently, CNTs suspended with sodium dodecylbenzenesulfonate (SDBS) surfactant have been deposited directly onto paper to form a porous, conductive CNT electrode [9–10]. Hu et al., were able to show a ~3X increase in specific capacitance when they compared CNT films deposited onto Xerox paper substrates to CNT films deposited onto PET plastic substrates [9]. This technique was selected as a starting point for our supercapacitor electrode fabrication due to its apparent performance advantage when compared to other published techniques.

2. CNT Supercapacitor Electrode Fabrication and Testing

In order to realize a packaged flexible supercapacitor, it was necessary to develop an electrode fabrication technique. Standard copier paper was chosen as a substrate, and a variety of CNT suspensions were deposited onto the substrate by drop casting.

A hot plate was used to heat 1 cm^2 substrates, which sped up the drop casting process by aiding in solution evaporation. Hot plate temperatures ranged from 80 °C to 150 °C depending on the CNT solution being deposited. Typically, 25 □l of CNT solution was placed on the substrate and allowed to sit until the paper dried. The process was repeated until the desired CNT solution volume was deposited onto the paper.

CNT/paper electrodes were evaluated both by their CNT film morphology and their electrochemical performance as supercapacitor electrodes. CNT film morphology was analyzed using an FEI Quanta environmental scanning electron microscope (ESEM). After CNT deposition was complete, electrodes were evaluated in the ESEM without further preparation. Following ESEM analysis, the electrodes were prepared for electrochemical analysis by clamping one edge of the electrode with nickel foil in order to provide an inert connection to the measurement apparatus (figure 1).

A second piece of Ni foil was used to clamp the electrode to an acrylic plate, allowing the entire electrode to be submerged in a liquid electrolyte for testing.

before and after CNT deposition. However, because of variations in the water content of the paper substrate, it was impossible to obtain an accurate measurement of the deposited CNT mass. The densities of the CNT solutions were confirmed by dispensing a known volume of CNT solution onto an aluminum foil sheet and weighing the sheet before and after deposition.

The resistance values of the CNT films covering the paper substrates were measured using a digital ohmmeter. The paper was probed repeatedly at different locations, which is why a range of resistance values appears. This is not intended to be a quantitative measure of the true sheet resistance of each electrode, but rather a comparative measure between the five samples above. Specific capacitance values were calculated from cyclic voltammetry measurements described above. 1M KOH was used as the liquid electrolyte, and scan rates of 20 mV/s were used for all five sample measurements.

As can be seen in table 1, the specific capacitance values achieved by the five electrode samples vary widely. The surfactant-free COOH SWCNTs in water clearly outperform the other CNT solutions, reaching a specific capacitance of 91.25 F/g. This value is on par with the best CNT supercapacitor electrodes that do not use pesudocapacitive materials. We believe that the surfactants and dispersants that are used to suspend the CNTs in the other solutions are having a detrimental effect on the electrode performance. This theory is supported by the CNT film morphologies observed under ESEM analysis in figure 2, where CNT bundling and contamination are clearly present in samples 1 and 4. CNT bundling was also observed in sample 5, but the severity of the bundling was clearly less than the other samples.

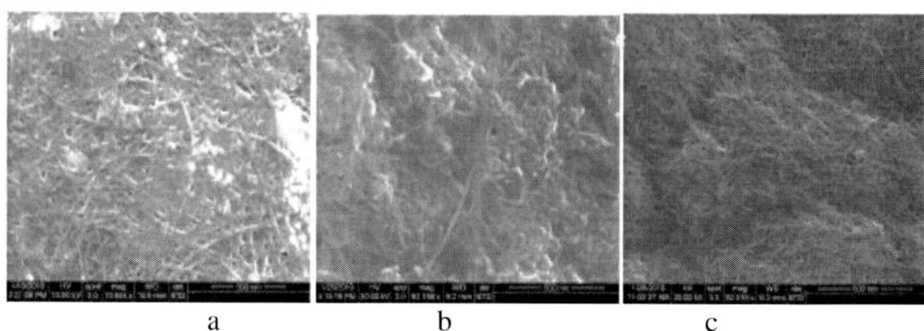

Figure 2. ESEM images of COOH SWCNTs suspended in acetone (a) (sample 1), SWCNTs in SDBS/diH2O (b) (sample 4), and COOH SWCNTs in H2O (c) (sample 5), deposited on paper substrates. Scale bars = 500 nm.

There were also no signs of the additional contamination that was observed in samples 1 and 4. Samples 2 and 3 (not shown) also contained CNT film contamination similar to samples 1 and 4. It is likely that the open pore structure seen in sample 5 allows electrolyte ions to more completely penetrate the CNT film, resulting in higher specific capacitance.

2.2. Impact of Surfactant on Performance

The use of surfactants and dispersants in the CNT solutions appear to have a negative impact on the overall device performance, resulting in excessive CNT contamination. We were unable to achieve the ~200 F/g specific capacitance described by Hu et al., using their SWCNT/SDBS solution protocol [9]. It is possible that differences in the starting CNT materials resulted in the differences in specific capacitance seen between sample 4, 5, and Hu's work. In order to confirm the impact of SDBS surfactant on electrode performance, a new CNT solution was made. 10 mg SDBS powder was mixed with 2 ml of the COOH SWCNT in water suspension used to fabricate sample 5. This eliminated the variability between nanotubes used, and should provide an accurate measure of the impact of having SDBS surfactant present in the CNT solution. Sample 8 was made by drop casting 2 ml of the new solution onto a paper substrate.

CNT film morphology was clearly impacted by the addition of the SDBS surfactant to the COOH SWCNT/water solution. The CNT film porosity has been visibly degraded, leading to reductions in the discharge current of the CV plots in figure 3. Table 2 shows a 10X reduction in specific capacitance due to the surfactant addition, with a large increase in resistance compared to sample 5. This indicates that the CNT conducting network is degraded, and the surface area available for the electrolyte ions to access has been reduced.

After device testing was complete, sample 8 was soaked in a deionized water bath for 30 min in an attempt to wash away unwanted surfactant molecules. We then loaded the sample into a furnace and heated to 180 °C for 2.5 hr. Additional device testing was performed, and the results are shown in table 2 under sample 8*. The washing and heat treatment did improve the electrode performance, most likely by driving off some of the surfactant, but electrode performance was still 3X below the sample that did not contain surfactant. Further rinsing and heat treatments did not improve device performance. The results of this study indicate that surfactants should be avoided in order to preserve CNT film porosity and to maximize electrode conductance and capacitance.

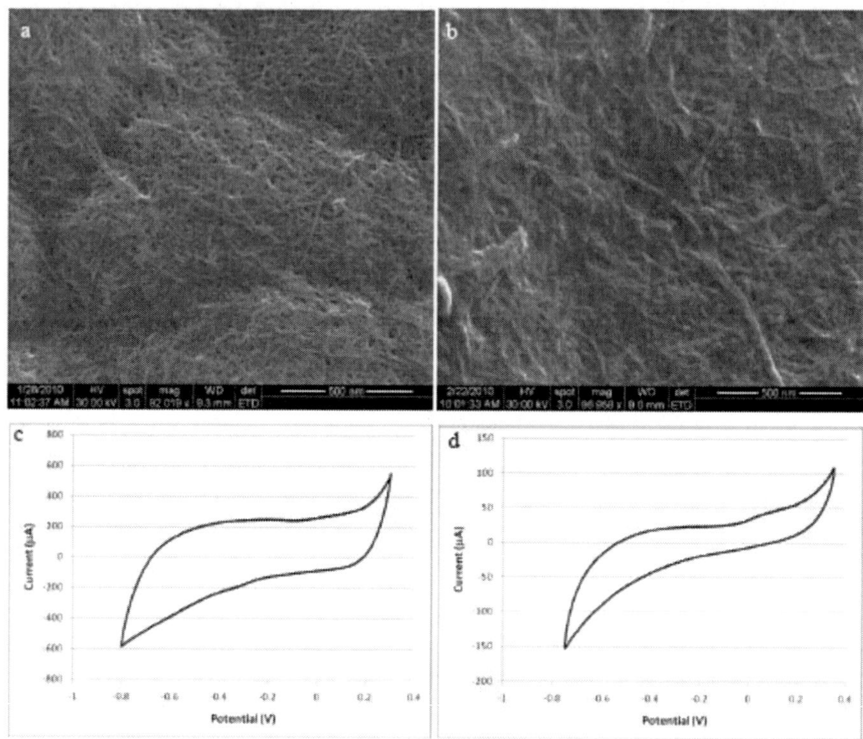

Figure 3. CNT film morphology and CV plots of SWCNT supercapacitor electrodes both with and without SDBS surfactant. ESEM images show (a) sample 5 with no SDBS, and (b) sample 8 with SDBS. CVs were taken at a scan rate of 20 mV/s and a scan range of –0.7 V to +0.3 V for both (c) sample 5, and (d) sample 8. ESEM scale bars = 500 nm.

Table 2. Comparison of electrodes fabricated with and without SDBS surfactant. Sample 8* has undergone rinsing and heat treatment to remove unwanted surfactant molecules

Sample	CNT type	Solvent	Volume Deposited/ CNT mass Deposited	Resistance	Specific Capacitance
5	COOH SWCNT	water	2 ml/0.1 mg	150–200 Ohm	91.25 F/g
8	COOH SWCNT	SDBS	2 ml/0.1 mg	50–200 kOhm	10.9 F/g
8*	COOH SWCNT	SDBS	2 ml/0.1 mg	0.8–1.5 kOhm	28.8 F/g

2.3. CNT Mass Loading Study

In order to ascertain if the entire CNT film was being accessed by the electrolyte, we performed a mass loading study. Three electrodes (samples 5–7) were fabricated using COOH SWCNTs in water, as described in table 3. The only variable changed in this experiment was the volume of CNT solution that was deposited onto the paper substrates. If the electrolyte is not able to penetrate the CNT film, the specific capacitance would be reduced by roughly 50% for each doubling of overall CNT mass because the added mass would not increase the overall capacitance of the electrode. CVs, shown in figure 4, were captured for all three samples under the same experimental conditions. Again, 1M KOH was used as the electrolyte, and a scan rate of 20 mV/s was used to generate the plots. A scan range of –0.8 V to +0.3 V was used for all three scans. The roughly rectangular shape of the CV plots is indicative of capacitive charge storage. As seen in table 3, no clear trend in specific capacitance was observed when the CNT mass loading was increased. The differences in specific capacitance between the three samples may have been due to variabilities in the drop casting process, which is done by hand and does not ensure consistent CNT film coverage across the entire substrate. However, it is clear than the liquid electrolyte is able to penetrate the varying thicknesses of the CNT films of samples 6–8. If this was not the case, a clear downward trend in specific capacitance would have been observed.

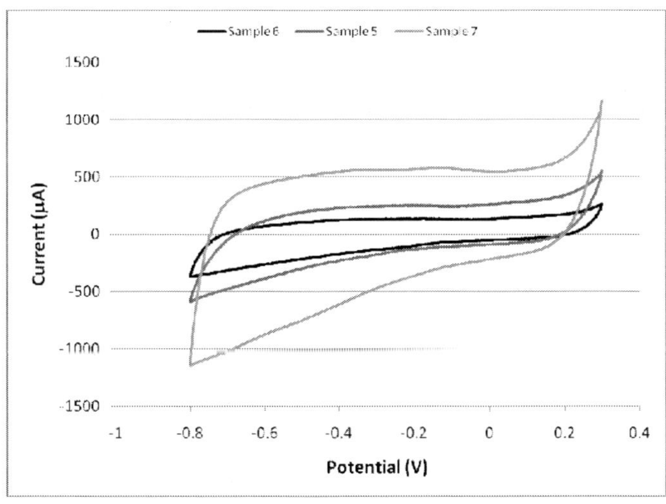

Figure 4. Cyclic voltammograms of sample 5, sample 6, and sample 7 using a 20 mV/s scan rate.

Table 3. Impact of mass loading on CNT/paper electrode performance

Sample	CNT type	Solvent	Deposited Volume/ CNT mass Deposited	Resistance	Specific Capacitance
6	COOH SWCNT	water	1 ml/0.05 mg	100–300 Ohm	101.5 F/g
5	COOH SWCNT	water	2 ml/0.1 mg	150–200 Ohm	91.25 F/g
7	COOH SWCNT	water	4 ml/0.2 mg	80–160 Ohm	115.6 F/g

2.4. MWCNT Electrodes

In addition to SWCNT electrodes, several multi-walled CNT (MWCNT) electrodes were made. It is expected that the specific capacitance of MWCNT electrodes would be less than a comparable electrode made using SWCNTs because the electrolyte ions cannot access the inner shells of the MWCNTs. This means that the inner tubes of a MWCNT would add weight to the electrode without providing any additional capacitance. MWCNT electrodes were expected to be more conductive due to the fact that MWCNTs are metallic, whereas SWCNT solutions contain a mixture of semiconducting and metallic CNTs. A commercially available MWCNT suspension in water with no added surfactants or dispersants was used to fabricate the MWCNT/paper electrodes.

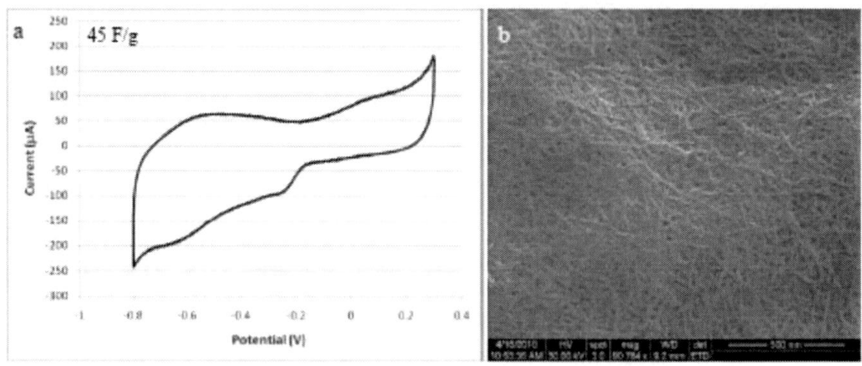

Figure 5. CV (a) and ESEM image (b) of a MWCNT/paper supercapacitor electrode. ESEM scale bar = 500 nm.

1 ml of 0.05 mg/ml MWCNT solution was drop cast onto a paper substrate that was placed on a 150 °C hot plate. The sample was then imaged, and finally tested in the electrochemical half cell measurement setup with the results shown in figure 5.

The electrode CNT morphology appears similar to the SWCNT electrodes. An open pore structure is maintained and no obvious signs of contamination are present. The electrode had a specific capacitance of 45 F/g, roughly half of the capacitance of a SWCNT electrode. The CV plot does show several peaks that are more pronounced than observed in the SWCNT electrodes. These may be due to surface groups on the MWCNTs that were not present on the SWCNT devices. The specific capacitance of this electrode was evaluated at a potential on the CV plot that was free of obvious redox peaks that could confuse the results (–0.15V was used for this electrode). We observed a resistance of 50–150 ohms which is lower than the typical SWCNT electrode.

3. CNT Flexible Supercapacitor Device Assembly

Following the CNT/paper electrode development work, we considered the issue of electrolyte selection and packaging. Liquid electrolytes are advantageous for many systems due to their superior ion mobility when compared to solid or gel electrolytes. However, liquid electrolytes are more prone to leaking when used in a flexible device. Electrolyte leakage not only compromises the functionality of the device, but could cause damage to its surrounding environment. In order to address these shortcomings, the performance of liquid KOH electrolytes was compared to a PVA:KOH gel electrolyte.

For liquid KOH supercapacitors, a 1M KOH solution was used. The PVA:KOH gel electrolyte was prepared in a ratio of 3:1 PVA to KOH [13]. To begin, a 70 mg/ml solution of PVA in deionized water was made. The solution was heated to a temperature of 130 °C and stirred continuously until the liquid turned from opaque to clear. Following the solution clearing, the appropriate amount of 1M KOH was added and allowed to mix for 5 min to achieve the proper ratio. The PVA:KOH solution was then poured into PTFE dishes and allowed to dry into a moist gel. The result is a very robust opaque white elastomeric gel. Care must be taken not to allow the gels to dry completely, as they lose flexibility as well as their ability to perform as an electrolyte.

The material chosen to encase the electrodes and electrolyte was a flexible polyester film. The film was sealed using a thermal sealer to melt the polyester together and form a pouch that encapsulates the device. The film did not degrade in the presence of KOH. Assembling full devices also required a new method for making electrical contact with the electrodes. To accomplish this, the same Ni foil clip was applied to one edge of the electrode. A thin Ni wire was then tack welded to the Ni foil clip. This provided good electrical contact, and also allowed the polyester pouch to be sealed around the thin wire to limit electrolyte leakage.

Three packaged supercapacitor structures were fabricated and compared, and are represented schematically in figure 6. The liquid KOH electrolyte architecture used 1M KOH as the electrolyte. Two CNT/paper electrodes were arranged so that the CNT films were facing each other. The two electrodes were separated by a third piece of plain copy paper that had been soaked in the KOH solution. The stack was assembled in a polyester pouch, which was then thermally sealed around the device. The gel electrolyte sandwich architecture used a PVA:KOH gel electrolyte. Again, two CNT/paper electrodes were assembled facing each other. A piece of PVA:KOH gel was cut to the appropriate size and placed between the two electrodes. This gel served as both the electrolyte and the separator that prevented electrode shorting. Again, the electrode/electrolyte stack was sealed in a polyester casing. The cast gel electrolyte architecture involved casting the electrolyte gel over and around CNT/paper electrodes, encapsulating the individual electrodes. This was done by placing each electrode in a small dish, and pouring hot liquid PVA:KOH solution over the electrodes. The gel was allowed to form, encapsulating the electrodes. Two gel encapsulated electrodes were then sandwiched together without any separator, and sealed in the package. We hypothesized that the gel casting process would allow the gel electrolyte to more completely penetrate the CNT film, proving superior capacitance and increased performance stability under bending when compared to the gel sandwich structure.

One additional issue that arose when assembling the packaged supercapacitors was non-flat CNT/paper electrodes. During CNT deposition, it was common for the paper to become wrinkled due to the repeated wetting and drying that it was subjected to. This became a problem when assembling packaged devices. It also made casting electrodes in the gel electrolyte very difficult because it was necessary to completely cover the electrodes with the gel in order to prevent device shorting. To address these issues, the CNT/paper electrodes were flattened prior to being fitted with the Ni clips and assembled.

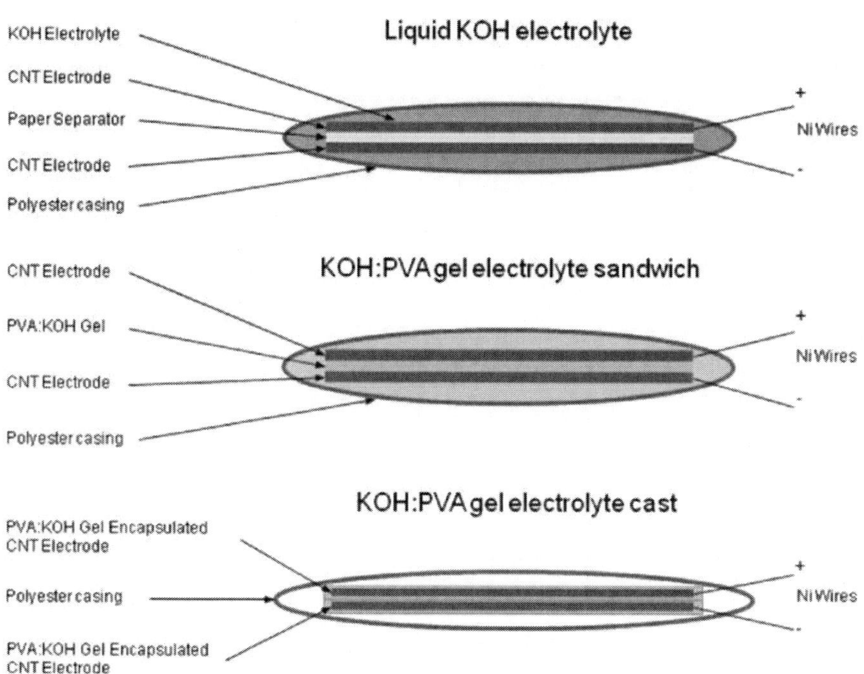

Figure 6. Full cell packaged flexible supercapacitor architectures.

The electrodes were first soaked in deionized water for 1 min to wet the substrates. The electrodes were removed from the water bath and sandwiched between a piece of absorbent cleanroom paper and a thick glass microscope slide. A small amount of weight was placed on top of the glass slide, forcing the paper completely flat. The electrodes were left in this configuration for several hours until the substrates were dry. This method was successful in producing very flat CNT/paper electrodes, which made device assembly much easier.

4. CNT Flexible Supercapacitor Device Testing

Device testing began by constructing three packaged devices. One of each device configuration described above was constructed using MWCNT/paper electrodes as described in section 2.4. Assembled devices were tested in a full cell configuration. All CVs were taken at 20 mV/s scan rates. Note that an assembled full cell supercapacitor is actually two capacitors in series, where the total capacitance $C_t=(C_1 \times C_2)/(C_1 + C_2)$. The devices were tested both flat and under static bending with the results shown in figure 7. Static bend tests were performed by wrapping the packaged devices around mandrels of varying diameters and taking CV measurements under each bending condition. We used rubber bands to secure the supercapacitors to the mandrels. A pressure test was also conducted on each device, where a small amount of weight was placed on top of the supercapacitor package, and CV measurements was taken under this condition.

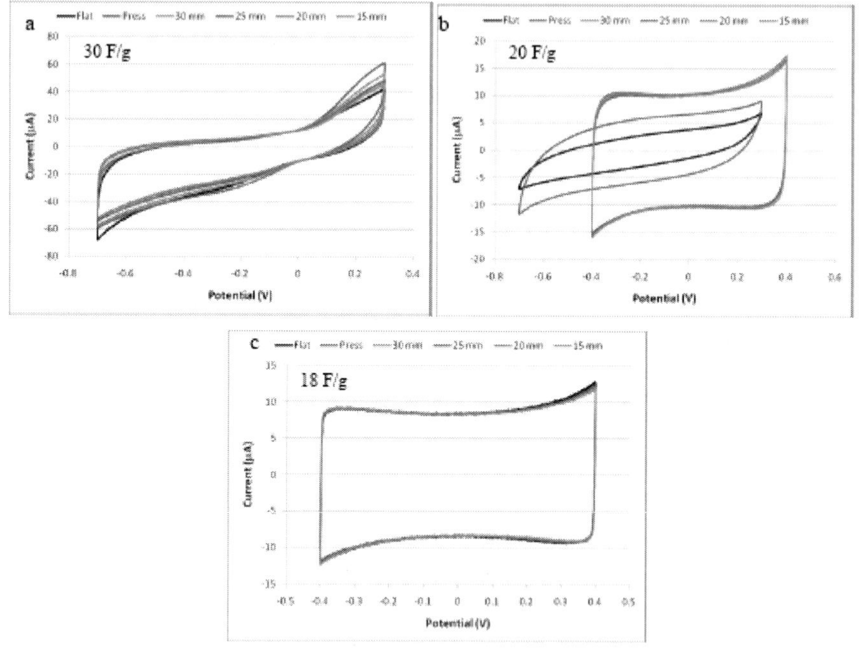

Figure 7. Cyclic voltammagrams of (a) Liquid KOH electrolyte packaged supercapacitor, (b) PVA:KOH gel sandwich packaged supercapacitor, and (c) PVA:KOH cast gel packaged supercapacitor under varying states of bending.

The liquid electrolyte device showed the best energy storage performance, with a specific capacitance of 30 F/g. This was expected, as the KOH ions are much more mobile in the liquid than they are in the gel matrix. There was some variability as the bending conditions were changed, but no appreciable trend was observed. The gel sandwich supercapacitor showed greater performance variability due to the loading conditions on the device. The flat or unloaded test showed very little capacitance. Applying pressure to the device increased the capacitance, as did the bend tests. No appreciable variability was observed between the four different bend diameters. It is believed that the pressure applied during the pressure and bending tests forced the PVA:KOH gel electrolyte into more intimate contact with the electrodes, improving ion intercalation and energy storage. It is also possible that liquid KOH is released from the gel during bending, which wet the electrodes and improved device performance.

The gel cast supercapacitor showed the lowest specific capacitance (18 F/g), but showed no sensitivity to its loading conditions. The lower capacitance observed in the gel cast supercapacitor may be due to PVA molecules occupying some of the CNT surface area, making them inaccessible to KOH ions.

CONCLUSION

Flexible packaged supercapacitor devices have been successfully fabricated and tested. Electrode fabrication studies showed a strong correlation between CNT film cleanliness and electrode performance. Dispersant agents used to suspend CNTs in solution proved detrimental to electrode performance, resulting in low specific capacitance values. Electrodes fabricated using commercially available dispersant-free CNT suspensions in water from Brewer Scientific proved to produce the best performing CNT/paper electrodes, with specific capacitance values in excess of 100 F/g.

Flexible full cell capacitors were produced in three different configurations. Each packaged device architecture produced unique device performance.

The liquid electrolyte device produced the highest specific capacitance, but the possibility of electrolyte leakage is a major disadvantage of this device design. The gel sandwich supercapacitor showed the second best device performance under bending, but its performance was variable under different bending conditions. This variability is undesirable, as it would introduce changes in energy storage capability under different system loading conditions.

The gel cast supercapacitor had the lowest specific capacitance of the three assembled devices.

However, the fact that device performance did not vary under any of the bending conditions, coupled with the low likelihood of electrolyte leakage when compared to a liquid electrolyte device, makes the gel cast device architecture attractive despite the performance penalty.

REFERENCES

[1] Pandolfo, A. G.; Hollenkamp, A. F. J. Power Sources 2006, 157, 11–27.

[2] Reddy, A. L.; Amitha, F. E.; Jafri, I.; Ramaprabhu, S. *Nano Res. Lett.* 2008, 3, 145–151.

[3] Amitha, F. E.; Reddy, A. L.; Ramaprabhu, S. *J. Nanopart Res.* 2009, 11, 725–729.

[4] Chen, P.; Shen, G.; Sukcharoenchoke, S.; Zhou, C. *Appl. Phys. Lett.* 2009, 94, 043113.

[5] Kaempgen, M.; Chan, C.; Ma, J.; Cui, Y.; Gruner, G. *Nano Lett.* 2009, 9, 1872–1876.

[6] Wang, D.; Song, P.; Liu, C.; Wu, W.; Fan, S. *Nanotechnology* 2008, 19, 075609.

[7] Ci, L.; Manikoth, S.; Li, X.; Vajtai, R.; Ajayan, P. *Adv. Mater.* 2007, 19, 3300–3303.

[8] Zhao, X.; Tian, H.; Zhu, M.; Tian, K.; Wang, J. J.; Kang, F.; R. Outlaw, A. *J. Power Sources* 2009, 194, 1208–1212.

[9] Hu, L.; Choi, J. W.; Yang, Y.; Jeong, S.; La Mantia, F.; Cui, L.; Cui, Y. PNAS Early Edition 2009.

[10] Hu, L.; Pasta, M.; La Mantia, F.; Cui, L.; Jeong, S.; Deshazer, H.; Choi, J.; Han, S.; Cui, Y. *Nano Lett.* 2010, 10, 708–714.

[11] Isvan, R. The 19th Int'l Seminar on Double Layer Capacitors and Hybrid Energy Storage Devices, Deerfield Beach, FL, 7–9 Dec 2009.

[12] Islan, M.; Rojas, E.; Bergey, D.; Johnson, A.; Yodh, A. *Nano Lett.* 2003, 3, 269–273.

[13] Lewandowski, A.; Skorupska, K.; Malinska, *J. Solid State Ionics* 2000, 133, 265.

In: Carbon Nanotubes
Editor: Percy Szalkowski
ISBN: 978-1-62618-493-0

Chapter 3

IMPROVING MICROBOLOMETRIC RESPONSE USING CARBON NANOTUBES*

Nichelle Perera and Priyalal Wijewarnasuriya

LIST OF SYMBOLS, ABBREVIATIONS, AND ACRONYMS

CNT	carbon nanotubes
CVD	chemical vapor deposition
I-V	current-voltage
SEM	scanning electron
SWCNT	single-walled carbon nanotubes
TCR	temperature coefficient of resistance
VO_x	vanadium oxide

ABSTRACT

Single-walled carbon nanotubes (CNTs) provide a semiconductor material with high sensitivity, detectivity, and a high temperature coefficient of resistance (TCR), in comparison to previously used vanadium oxide microbolometers, to detect infrared radiation. The fabrication of the microbolometer uses photolithography and metal

* This is an edited, reformatted and augmented version of an Army Research Laboratory publication, ARL-TN-0523, dated January 2013.

deposition processes to create electrical contact pads on the nanotubes, and mounting and wire bonding to a leadless chip carrier. Once fabricated, an experimental dewar is used to test the device at varying temperatures, ranging from 78 K to room temperature. The results indicate that using a semiconductor material with a higher TCR provides a wider range of electrical resistance, and improves the sensitivity and response of infrared imaging technology.

1. Introduction/Background

The ability to detect and interpret the incoming infrared radiation of a scene or object that is not visible (to the naked eye) is essential for the Army to maintain the upper hand. To be better aware of the surroundings, infrared detectors, which can remotely measure the temperature of an object or a scene, require microbolometers with high sensitivity, high detectivity, and a large bolometric response [6]. Such microbolometers will produce inexpensive, uncooled infrared detectors. To improve the bolometric response of infrared sensors, single-walled carbon nanotubes (SWCNT) can be used as the semiconductor material to absorb the radiation. In addition to detecting radiation, these CNT-based sensors have potential for different applications, such as detecting agents other than radiation. For example, CNT-based sensors can also detect chemical and biological agents, and thus, has potential use in other areas [6]. For these sensors, individual CNTs showed promise as they can detect non-visible light.

As a solution, a uniform system of nanotubes provides an easily mass-producible material. Figure 1 shows a system of SWCNT deposited onto a sapphire substrate. In comparison to more commonly used vanadium oxide (VO_x) microbolometers, CNT-based bolometers provide better resistivity, as well as higher sensitivity and detectivity. Incoming infrared radiation is absorbed by the SWCNT, raising its temperature. The thermal response is associated with the electrical response, i.e., the temperature change leads to a change in resistance.

The measured change in resistance, which results from the change in temperature, corresponds to temperature values, which are represented in the resulting infrared image [3, 4]. CNTs provide a wider range of resistance values, dependent on the temperature.

A larger variation provides greater clarity in the infrared image by increasing the temperature resolution.

Figure 1. Scanning electron microscopy (SEM) picture of spray-deposited CNT on sapphire substrate.

CNT-based microbolometers offer multiple advantages over previously used VOx microbolometers. CNTs can provide temperature coefficient of resistance (TCR) values as high as those obtained from VOx, but CNT films are more easily reproduced. There is experimentation being done to obtain TCR values even higher than VOx, when certain conditions are modified to increase sensitivity. For other factors, such as detectivity, CNT-based experiments result in values comparable or better than those obtained from VOx microbolometers, as shown by table 1 [6].

Table 1. Comparisons of performance metrics between VOx and CNT-based microbolometers

Performance Metrics	Current VOx Film Based	Expected From the Proposed CNT Technology
Frame rates	30–60 Hz	100–1000 Hz
Sensitivity (TCR)	0.02 to 0.025/K	≥ (0.01–0.025/K)
Pixel size	25 μm	<25 μm
Detectivity, D*	"(4–7) x 109 cm Hz 1/2	"(4–7) x 109 cm Hz 1/2 W–1
Compatible with printed electronics requirements	No	Yes

SWCNTs have a higher TCR, as well, which leads to higher sensitivity [5, 6].

TCR is defined as the change in resistance per Kelvin, divided by the resistance measured at room temperature [5]

$$TCR = \frac{1}{R_e}\frac{dR_e}{dT} \quad (1)$$

2. EXPERIMENT/CALCULATIONS

To create the system of SWCNTs, the chemical vapor deposition (CVD) method was used. A carrier gas, consisting of argon and hydrogen, acted as the carbon source. The chips were heated to decompose the gases into the SWCNT network. Once the SWCNTs were situated on top of silica oxide, which sits upon the sapphire substrate, photolithography was performed on the samples to define the contact regions. First, resist was spun onto the samples at 3500 rpm and then baked at 75 °C. The samples were then exposed for 4 s and post-expose baked. Once the samples were flood exposed and developed, the areas for the metal to be deposited were defined within the remaining resist. Figure 2 provides a schematic of the SWCNT fabrication process.

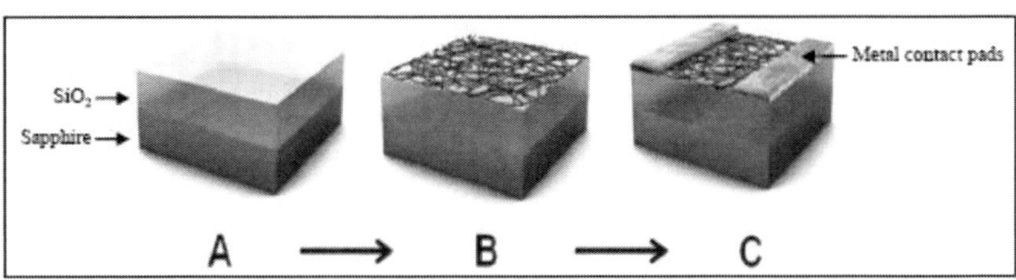

Figure 2. A schematic of the SWCNT fabrication process; (a) substrate template (sapphire/SiO2), (b) addition of SWCNTs, and (c) deposition of metal contact pads.

Two metals, titanium and gold, were evaporated and deposited onto the samples. First, the pressure was pumped down to 1.9×10^{-6} Torr. Then, 0.15 nm of titanium was deposited, followed by 2.5 nm of gold. Once deposited, acetone was used to lift off the remaining resist. The removal of the resist also removed the excess metal surrounding the contact pads. Once complete, the samples were left with metal contact pads sitting atop the system of SWCNTs.

After fabricating the samples, they were wire bonded onto a 68-pin leadless chip carrier (figure 3). Once bonded, the device was measured for current and resistance versus voltage at temperatures ranging from 78 to 300 K. Liquid nitrogen, which flows to the dewar, was used to cool the chip. A parameter analyzer, 4156C, took current-voltage (I-V) measurements at varying temperatures. The range of voltages tested was from –2.0 to 2.0 V.

Using Ohm's law, resistance was calculated for each device at each temperature.

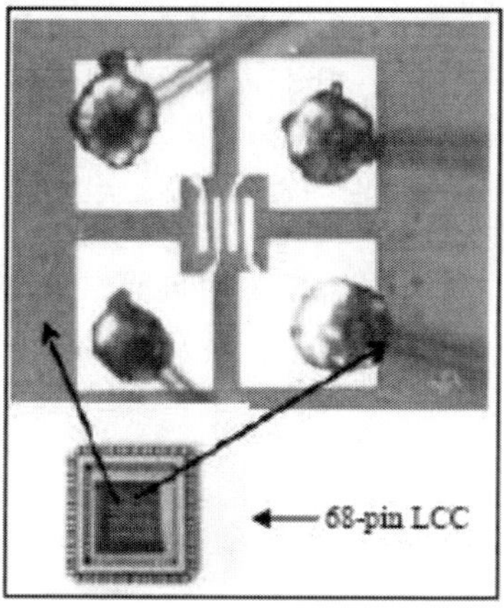

Figure 3. The 68-pin leadless chip carrier: wire-bonded metal contact pads on CNT network.

3. Results and Discussion

As mentioned, I-V measurements were taken at temperatures ranging from 78 to 300 K, at a voltage range of –2 to 2 V. Figure 4 shows the I-V measurements for a device at 78, 105, and 300 K. The current is seen to increase as the temperature increases, as expected.

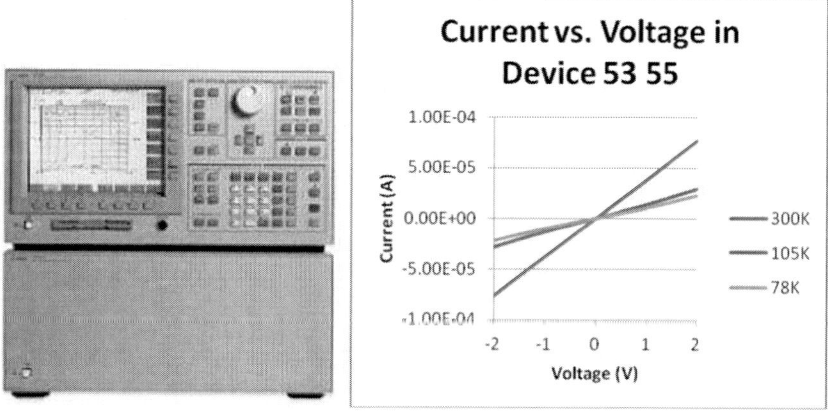

Figure 4. (a) (left) Parameter analyzer 4156C and (b) (right) measured current-voltage vs. temperature using 4156C.

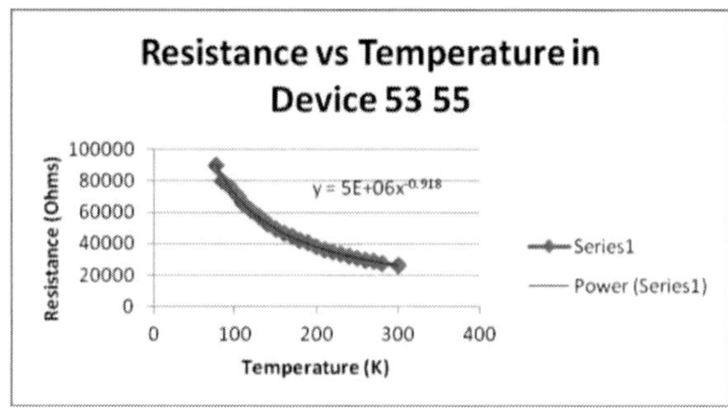

Figure 5. Resistance vs. temperature graph: As temperature decreases, resistance increases, providing a TCR of ~3.12% at 300 K.

Figure 5 shows that as temperature increases from 78 to 300 K, the resistance decreases. Using the aforementioned equation, the TCR is calculated to be ~3.12% at 300 K.

Conclusion

The fabrication of a microbolometer using SWCNTs, which includes the processes of chemical vapor deposition, photolithography, metal deposition, and wire bonding, creates a microbolometer with a better ability to detect incoming infrared radiation. The SWCNT device exhibits a high TCR of 3.12% when exposed to temperatures varying from 78–300 K. This TCR is comparable to that calculated from VO_x microbolometers. Further experiments should be conducted using SWCNTs to measure the response to incoming infrared radiation with respect to frequency. Overall detectivity, D*, could also be examined in CNT-based sensors. By further experimenting and using SWCNT networks in microbolometers, infrared detectors could be improved such that they can be used for an even wider variety of applications in a greater number of situations.

References

[1] Eminoglu, S. Uncooled Infrared Focal Plane Arrays with Integrated Readout Circuitry Using MEMS and Standard CMOS Technologies. *Doctoral thesis.* The Middle Eastern Technical University, 2003.

[2] Hecht, J. *Photonic Frontiers: Room-Temperature IR Imaging:* Microbolometer Arrays Enable Uncooled Infrared Camera. Laser Focus World. April 2012. http://www.laserfocusworld.com/ articles /print/ volume-48/issue-04/features/microbolometerarrays-enable-uncooled-infrared-camera.html (accessed 7 August 2012).

[3] Liger, M. *Uncooled Carbon Microbolometer Imager. Doctoral thesis.* California Institute of Technology. CaltechTHESIS. http:// thesis.library.caltech.edu/3841/ (accessed 7 August 2012).

[4] Maine, S.; Koechlin, C.; Fleurier, R.; Haidar, R.; Bardou, N.; Dupuis, C.; Attal-Trétout, B.; Mérel, P.; Deschamps, J.; Loiseau, A.; Pelouard, J.-L. *Mid-infrared Detectors Based on Carbon Nanotube Films.* Phys. Status Solidi C 2010, 7, 2743–2746. http://dx.doi.org/10.1002/ pssc.200983827.

[5] Ashok K.; Sood, E.; Egerton, J.; Puri, Y. R.; Fernandes, G.; Kim, J. H.; Xu, J.; Goldsman, N.; Dhar, N. K.; Wijewarnasuriya, P. S.; Lineberry, B. I. *Carbon Nanotube Based Microbolometer, Development for IR Imager and Sensor Applications.* Proc. SPIE 2011, 8155, 815513. http://dx.doi.org/10.1117/12.895277.

[6] Cech, J.; Swaminathan, V.; Wijewarnasuriya, P.; Currano, L. J.; ovalskiy, A.; Jain, H. *Fabrication of Freestanding SWCNT Networks for Fast Microbolometric Focal Plane Array Sensor.* Proc. SPIE 2010, 7679, 76792N. http://dx.doi.org/10.1117/12.855589.

MWCNT	multi-wall carbon nanotube
Ni	nickel
NiCo	nickel cobalt
PLD	pulsed laser deposition
Pt	platinum
PVD	physical vapor deposition
RF	radio frequency
SEM	scanning electron microscopy
Si	silicon
SiO_2	silicon dioxide
VSM	vibrating sample magnetometer

ABSTRACT

The high demand for nanoelectronics devices, high density magnetic memories, sensors, and energy storage for the future force and warfighters have led us to develop unique methods to fill vertically aligned nanofibers with magnetic nanoparticles such as cobalt ferrite (CoFe2O4), iron (Fe), and nickel cobalt (NiCo). In this work, pulsed laser deposition (PLD) is used to fill CoFe2O4 and DC magnetron sputtering is used to fill Fe and NiCo in carbon nanotubes (CNTs). The filled CNTs are characterized by scanning electron microscopy (SEM), energy-dispersive spectrometry (EDS), atomic force microscopy (AFM), magnetic force microscopy (MFM), and vibrating sample magnetometer (VSM). Magnetization measurements in-plane and out-of-plane with respect to the sample's surface of CNTs filled with CoFe2O4 indicates a reasonable coercivity of 0.4 T. The magnetic anisotropy is, however, found out to be randomly oriented indicating a polycrystalline structure. The unique difference between the in-plane and out-of-plane magnetizations is the sharing produced by the demagnetizing field in the perpendicular direction.

1. INTRODUCTION AND BACKGROUND

Since the first comprehensive and detailed characterization of carbon nanotubes (CNTs) by Iijima in 1991 [1], the saga to fill the low-dimensional space inside CNTs with different types of applications-related materials has continued. Pristine CNTs possess incredible mechanical and electrical properties [2, 3] besides the hallowed enclosed volume of space. This low-dimensional space provides an excellent opportunity to enhance the property of CNTs by filling it with applications-related materials [4]. A few examples include improved photovoltaic performance demonstrated in platinum (Pt)-filled CNTs [5] and the possibility of lightweight wide-band microwave absorbers using ferromagnetic-filled CNTs [6]. CNTs filled with magnetic nanoparticles are attractive candidates for active elements in changeable diffraction gratings, filters, and polarizer [7].

Filling CNTs with both ferromagnetic and ferroelectric materials could be another way of coupling the two properties, hence fabricating multi-ferroic materials [8, 9]. Iodine-filled CNTs have shown high electrical conductivity and excellent mechanical property [10]. Molecular dynamic simulations [11–13] indicate improved mechanical behavior of filled CNTs compared to unfilled CNTs. Increased compressibility of iron (Fe)-filled CNTs has been observed with high-pressure x-ray diffraction [14]. The mechanical strength of filled CNTs and their buckling behavior make them ideal for use as tips for scanning probe microscopes [15]. CNTs filled with indium (In) can be used as nanoscale mass conveyers in a nano assembly tool [16].

When filling CNTs with conductive fluids, a voltage is generated between the tube ends indicating potential application of CNTs in flow sensing and converting mechanical energy to an electrical signal [17]. Silver-filled CNTs are promising to be used as spectroscopic enhancers and chemical sensors in the visible range [18]. Filling CNTs with hydrogen has application for mobile energy storage at room temperature [19].

The list of applications of nanotubes filled with magnetic material also includes materials for wearable electronics [20], cantilever tips in magnetic force microscope [21], magnetic stirrers in microfluidic devices, and magnetic valves in nanofluidic devices. Biomedical applications include capsules or nanosubmarines for magnetically guided drug delivery to desired locations in the body, and non-pervasive diagnosis and treatment that can bypass surgery [22].

Nanomagnets are also important from a fundamental point of view in understanding the physics of one-dimensional magnets. The low-dimensional enclosed space of CNTs offer an environment in which previously unknown physical phenomena at nanoscale level can be observed. For example, confinement of matter on the nanoscale can induce phase transitions not seen in bulk systems [23–27]. It can also be used as a template to synthesize nanowires and nanomagnets.

The major hurdle for applications-related filling is the assembly of ordered nanoscale structures and controlled fillings. There has been success in preparing aligned nanotubes vertically on silicon dioxide (SiO_2) substrate by chemical vapor deposition (CVD) technique [28]. In this work we report a single step procedure used to fill vertically aligned multi-walled (MW) CNTs with cobalt ferrite ($CoFe_2O_4$) using pulsed laser deposition (PLD) and with Fe using sputtering. This is the first attempt ever to fill CNTs using PLD [29] and magnetron sputtering [30, 31], which are the two commonly used techniques to prepare various applications related thin and thick films.

PLD is a type of physical vapor deposition (PVD) technique where a high power pulse laser beam is focused inside a vacuum chamber to strike a target of the material that is to be deposited. Thus, the material is vaporized from the target (in a plasma plume), which deposits itself as a thin film on a substrate (such as a silicon wafer facing the target). The removal of atoms from the bulk material is done by vaporization of the bulk at the surface region in a state of non-equilibrium. In this, the incident laser pulse penetrates into the surface of the material within the penetration depth. This dimension is dependent on the laser wavelength and the index of refraction of the target material at the applied laser wavelength (248 nm). The strong electrical field generated by the laser light is sufficiently strong to remove the electrons from the bulk material of the penetrated volume. The free electrons oscillate within the electromagnetic field of the laser light and can collide with the atoms of the bulk material, thus transferring some of their energy to the lattice of the target material within the surface region. The surface of the target is then heated up and the material is vaporized. In the second stage, the material expands in plasma parallel to the normal vector of the target surface towards the substrate due to Coulomb repulsion and recoils from the target surface. The spatial distribution of the plume is dependent on the background pressure inside the PLD chamber. PLD is only one of many thin-film deposition techniques. Other methods include sputter deposition (radio frequency [RF], magnetron, and ion beam).

Sputter deposition is also based on PVD in which materials are sputtered on a substrate from a target. Sputtered atoms ejected from the target have a wide energy distribution, typically up to tens of eV. The sputtered ions are ballistic from the target in straight lines and impact energetically on the substrates or vacuum chamber causing re-sputtering. Alternatively, at higher gas pressures, the ions collide with the gas atoms that act as a moderator and move diffusively, reaching the substrates or vacuum chamber wall and condensing after undergoing a random walk. The entire range from high-energy ballistic impact to low-energy thermal motion is accessible by changing the background gas pressure. The sputtering gas is often an inert gas such as argon. For efficient momentum transfer, the atomic weight of the sputtering gas should be close to the atomic weight of the target, so for sputtering light elements neon is preferable, while for heavy elements krypton or xenon are used. Reactive gases can also be used to sputter compounds. The compound can be formed on the target surface, in-flight or on the substrate depending on the process parameters.

The availability of many parameters that control sputter deposition makes it a complex process, but also allows experts a large degree of control over the growth and microstructure of the film. An important advantage of sputter deposition is that even materials with very high melting points are easily sputtered.

Sputtering sources are usually magnetrons that use strong electric and magnetic fields to trap electrons close to the surface of the magnetron, which is known as the target. The electrons follow helical paths around the magnetic field lines undergoing more ionizing collisions with gaseous neutrals near the target surface than would otherwise occur. The extra ions created as a result of these collisions lead to a higher deposition rate. It also means that the plasma can be sustained at a lower pressure. The sputtered atoms are neutrally charged, and hence, unaffected by the magnetic trap. Though, there have been previous attempts [32] to fill multi-walled carbon nantubes (MWCNTs) in aqueous suspension with $CoFe2O_4$, the methods presented in this work are more functionally oriented as the tubes are aligned and possess high symmetry.

2. MATERIALS AND METHODS

The vertically aligned MWCNTs used in this filling experiment are grown using a thermal CVD method [28] on a SiO_2 substrate. This method involves exposing silica structures to a mixture of ferrocene and xylene at 770 °C for 10 min.

The furnace is pumped down to ~200 mTorr in argon bleed and then heated to the temperature of 770 °C. The solution of ferrocene dissolved in xylene (~0.01g/ml) is pre-heated in a bubbler to 175 °C and then passed through the tube furnace. The furnace is then cooled down to room temperature. The open ended MWCNTs tips are filled with $CoFe_2O_4$ using PLD and with Fe and nickel cobalt (NiCo) using DC magnetron sputtering methods.

$CoFe_2O_4$ is used in this work due to its high anisotropy ($4x10^6$ ergs/cm^2) and magnetostriction ($800x10^{-6}$) at room temperature [33]. The target $CoFe_2O_4$ for laser oblation is prepared using the combination of ferric and cobalt oxides in a 2:1 ratio, respectively [34]. The polycrystalline $CoFe_2O_4$ filling in MWCNT is carried out in a high vacuum (2×10^{-7} Torr) with a 100-W pulsed excimer laser (krypton fluoride [KrF]) at 1.5 J/cm^2 and 3 Hz as the energy density and repetition rate at a 248-nm wavelength, respectively. During deposition, the SiO_2 template is heated at 300°C and the target is rotated in order to ensure its uniform wear. A total of 12,000 shots are fired to fill the nanotubes at 3 Hz in an oxygen pressure of 30 mTorr with output energy of 800 mJ.

DC magnetron sputtering is used to fill vertically aligned MWCNTs with NiCo and Fe. A gun power of 50 W and an argon pressure of 0.07 Torr are maintained during the 30-min deposition. Then, 51 nm of the sputtering material is deposited over the CNTs at the deposition rate of 1.7 Å/s. Deposition is carried out at an angle of 0° from the normal to the substrate surface. A second sample at an angle of 70° from the normal is also prepared. The latter deposition at 70° is an oblique deposition known for constructing nanocolumns. When deposition angle with the substrate normal is high, a columnar structure is observed [35]. The unique structure consisting of separated columns inclined toward the deposition direction is due to the "shadowing effect," which is caused by the height of the initial nuclei. A shadow is formed behind the initial nuclei where the rest of the vapor flux cannot reach. Since the surface mobility of the deposited adatoms is low at room temperature the growth rate in the normal to the surface is higher than that in the plane.

3. Results and Discussions

Scanning electron microscopy (SEM) of vertically aligned MWCNTs grown on a SiO_2 substrate before deposition is depicted in figure 1. As shown in the figure, most of the tubes are aligned vertically and only very few are misaligned. Filling by PLD method is mediated by the highly energetic particles in the plasma formed when an excimer laser of KrF of average power (100 W) and a wavelength of 248 nm strike the target material, i.e., $CoFe_2O_4$. Due to the high energy, in the order of several hundreds of eV's, the particles penetrate deep into the lumen of the CNTs. However, these highly energetic particles also increase the probability to damage the CNT openings by colliding with the edges. At present, the only mechanism to control this from occurring results in slowing all the particles by increasing background pressure.

This has a negative consequence, however. The particles do not have enough momentum to travel deep into the lumen of the CNT. In this work, 30 mTorr of background oxygen pressure is applied to prevent $CoFe_2O_4$ particles from colliding with the edges of CNTs.

Figure 2 depicts SEM of vertical tubes after being filled by PLD with $CoFe_2O_4$. From SEM measurements, it was not feasible to determine the filling depth; however, in single crystals, the depth can be determined using in- and out-of-plane magnetization measurements [36].

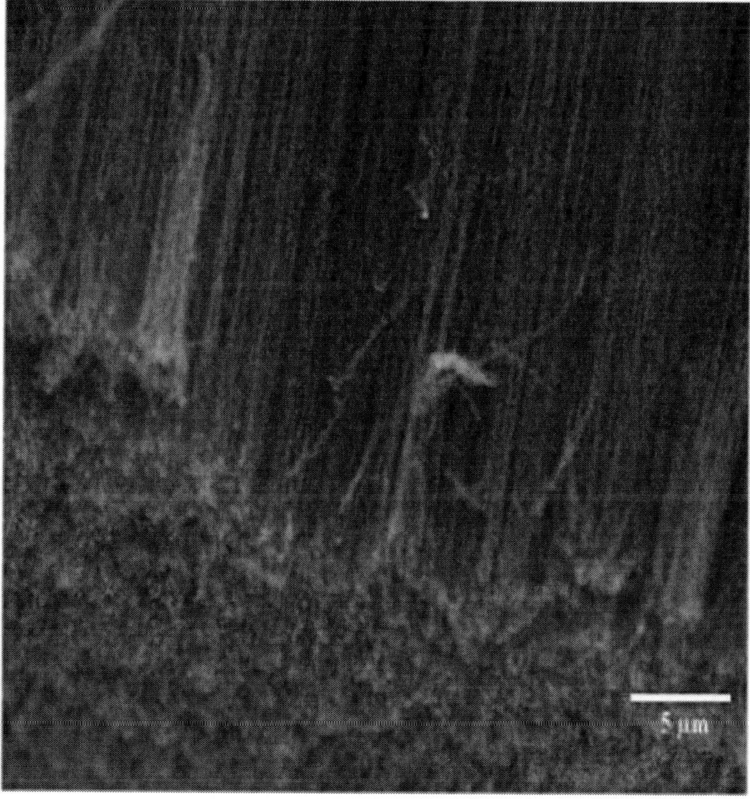

Figure 1. SEM of vertically grown MWCNTs on SiO_2.

Magnetization measurements are performed using vibrating sample magnetometer (VSM) both in-plane and out-of-plane loops with respect to the sample's surface. Figure 3 shows the magnetic moments with respect to the applied fields for the two loops, indicating the high coercivity of 0.4 T. The high coercivity is due to the coupling of the spins of the cobalt (Co) ions to the Fe ions in the oxide structure.

Figure 2. SEM of vertically grown MWCNTs on SiO_2 filled with $CoFe2O_4$ by PLD at high resolution (a) and lower resolution (b).

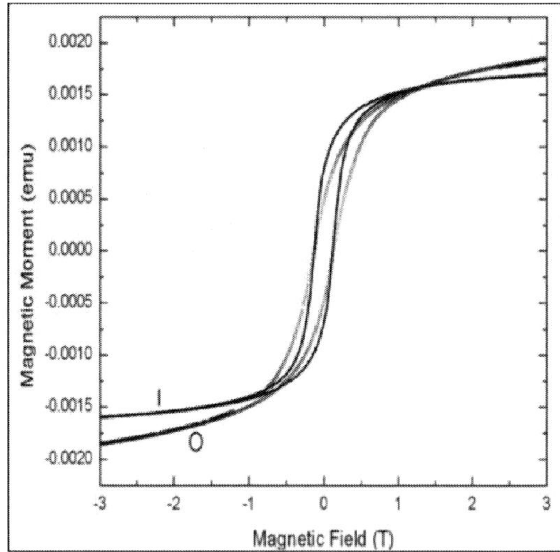

Figure 3. Magnetic moment vs. applied field; the solid line represent in-plane and the scatter line represents out-of-plane magnetization.

Magnetic properties of a material usually are very sensitive to its shape due to the dominating role of anisotropy in magnetism. In particular, the shape of the nanocrystals is a dominating factor for the coercivity. Such magnetic nanocrystals with distinct shapes possess tremendous potentials in technological applications of magnetic nanocrystals for high-density information storage and also in the fundamental understanding of magnetism. The coercivity of nanoparticles from Stoner-Wohlfarth theory is determined by anisotropy constant and saturation magnetization [37]:

$$H_C = 2K/(P_o M_S), \qquad (1)$$

where P_o is a universal constant of permeability in free space, K is the blocking temperature, and M_S is saturation magnetization.

The in-plane (solid line, figure 3) and out-of-plane (scattered line, figure 3) magnetization loops exhibit the same hysteresis. The in-plane and out-of-plane hysteresis curves coincide, indicating a high coupling between the in-plane and out-of plane anisotropy.

The slight difference between the in-plane and out-of-plane magnetizations is the sharing produced by the demagnetizing field in the perpendicular direction.

Figure 4 shows the SEM and energy dispersive spectroscopy (EDS) of vertically grown MWCNTs filled by DC magnetron sputtering with Fe. In figure 4a, the deposition is parallel to the normal of the substrate surface and in figure 4b the deposition is oblique at an angle of 70° to the normal of the substrate surface.

The EDS measurements show Fe peak and silicon (Si) peak which is due to the SiO2 substrate on which the CNTs are grown.

The graph of the magnetic moment of the Fe-filled MWCNTs as a function of applied field is shown in figure 5. Solid line represents in-plane and scatter represents out-of-plane magnetization.

In both cases, the in-plane and out-of-plane magnetization hysteresis loops coincide, indicating a strong coupling between the in-plane and out-of-plane anisotropy, which is mainly shape anisotropy due to the constricted spacing of the lumen of CNT. The hysteresis loop of Fe-filled MWCNT's deposited at 0° from the normal exhibits an anomalous narrowing of the loop at the zero magnetization axis.

This anomalous hysteresis loop indicates somewhat indifferent domain behavior and is subject to further investigation.

Figure 6 shows 5 x 5 μm^2 scan areas of atomic force microscopy (AFM) and magnetic force microscopy (MFM) images of Fe-filled MWCNTs at the same spot of the sample prepared by DC magnetron sputtering.

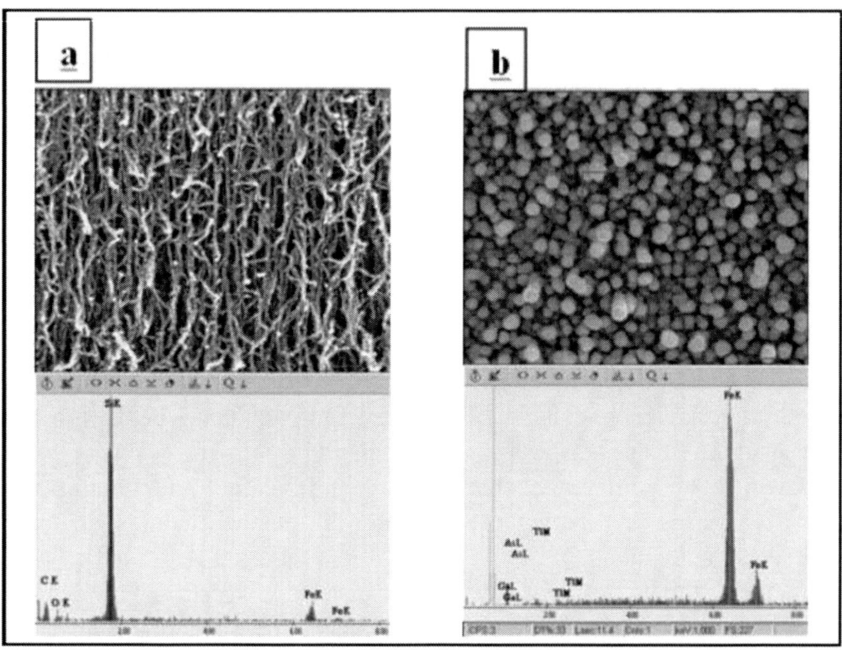

Figure 4. SEM of vertically grown MWCNTs on SiO_2 filled with Fe by DC magnetron sputtering at 0° from the normal of the substrate in (a) and at 70° from the normal (b).

The distinct bright and dark regions in the MFM images have a greater length scale than the corresponding atomic force morphology images. The average domain size *D* is determined by measuring the distance between successive bright to dark regions using line scans across the MFM image. The competition between magnetostatic, exchange, magnetocrystalline, and any other growth-induced anisotropy energies lead to stripe domain configurations in which the domain size depends on the sample thickness and the domain configuration depends on the sample magnetic history. The theoretical explanation of the formation of magnetic domains is to minimize the magnetic energy of a ferromagnetic crystal.

Figures 7 and 8 illustrate two-dimensional (2-D) and three-dimensional (3-D) versions of the 1.5 x 1.5 μm^2 scan area of magnified version of MFM image shown in figure 6b.

The stripe domains in both figures are clearly visible with typical domain width thickness of 0.25 μm.

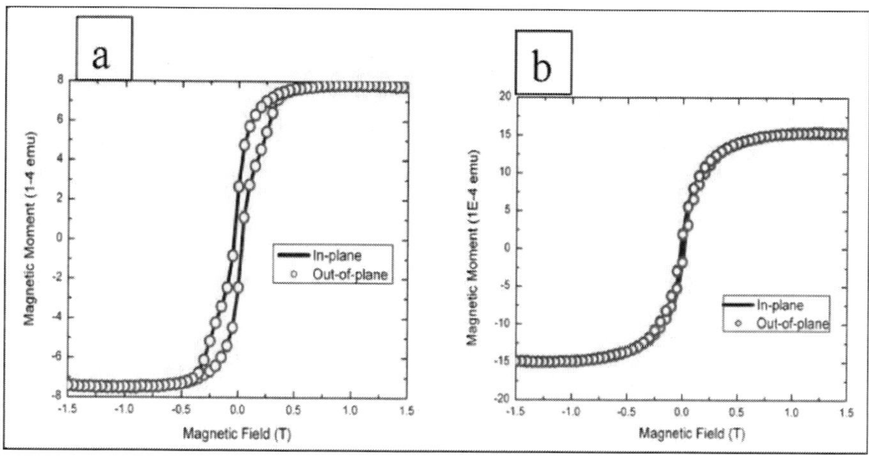

Figure 5. Magnetic moment vs. applied field; the solid line represent in plane and scatter represent out of plane magnetization of vertically grown MWCNTs on SiO_2 filled with Fe by DC magnetron sputtering at 0° from the normal of the substrate in (a) and at 70° from the normal (b).

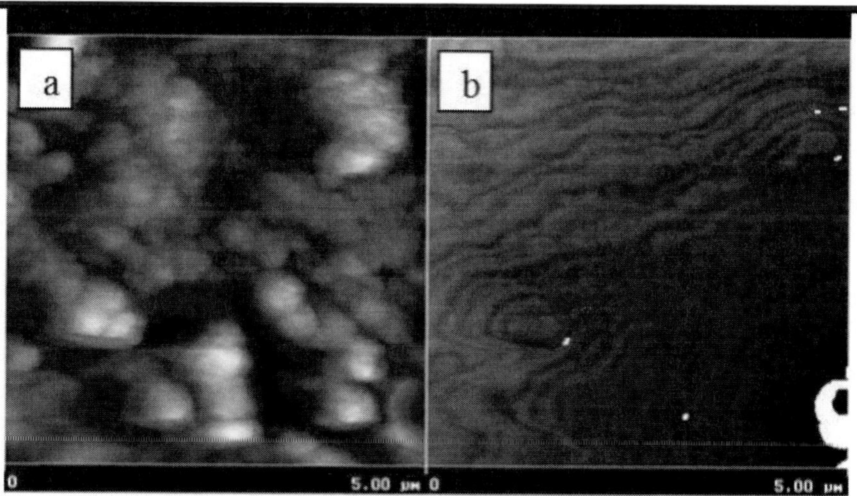

Figure 6. (a) AFM scan of MWCNTs filled with Fe sputtered head-on and (b) the corresponding MFM scan.

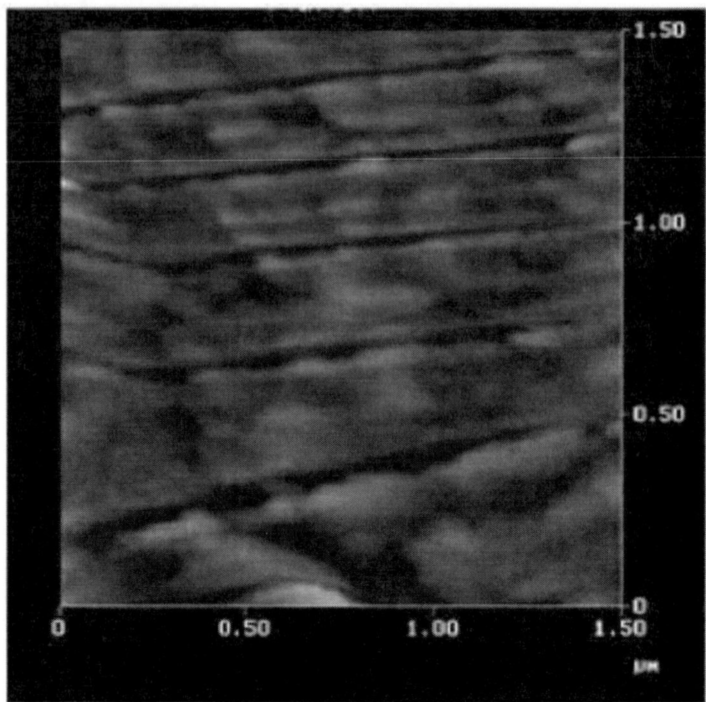

Figure 7. MFM scan of MWCNTs filled with Fe sputtered head-on.

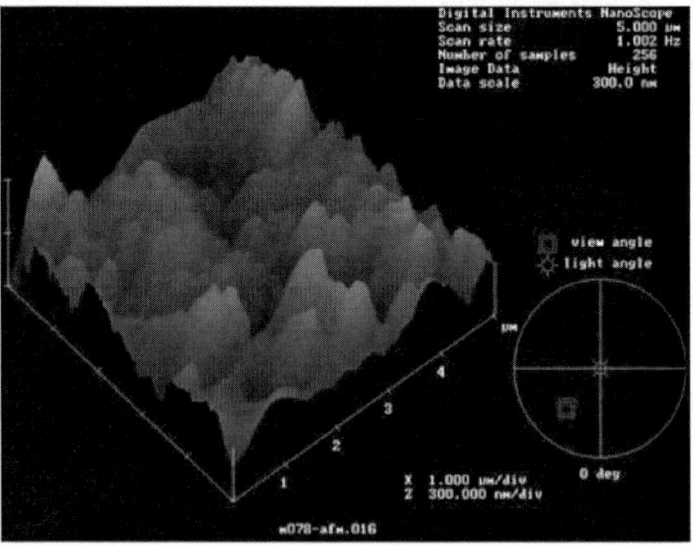

Figure 8. 3-D AFM scan of MWCNTs filled with Fe sputtered head-on.

CONCLUSION

In summary, for the first time, PLD technique and magnetron sputtering are used to fill vertically aligned CNTs with high-yield magnetic nanoparticles. The magnetization measurements suggest that the filled magnetic nanoparticles have a polycrystalline nature, as evidenced from the randomly oriented magnetic anisotropy.

Spin reorientation due to competition between shape anisotropy of nanotubes and magneto-crystalline component of anisotropy was observed while filling the MWCNTs with $CoFe_2O_4$, and seemed to be dependent on the lattice of the material.

The fillings of Fe and NiCo in the CNTs using magnetron sputtering method, showed an anomalous hysteresis in the VSM measurement and the stripe magnetic domains in the MFM measurement with an average thickness of 0.25 gm. This is indicative of a soft magnetic material.

We believe that our present work further extends the applications of CNT-based materials in nanoscale electronics technologies. This study opens up not only a simple and easy way preparation method and low-cost synthesis procedures, but also state-of-the-art industrial-scale preparation techniques. The fabrication of oxide nanowires using MWCNTs as templates will provide new approach to nanofabrication of ferromagnetic materials, high T_c superconductors, or colossal magnetoresistance (CMR) materials.

REFERENCES

[1] Iijima, S. Helical Microtubules of Graphitic Carbon, *Letters to Nature* 1991, 354, 56–58.

[2] Osborne, I. S. A Nanotube Story with a Twist, Science 2003, 302, 749.

[3] Tombler, W. T.; Zhou, C.; Kong, J.; Dai, H. Gating Individual Nanotubes and Crosses with Scanning Probes. App!. *Phys. Lett.* 2000, 76, 2412–2414.

[4] Ivanovskaya, V. V.; Köhler, C.; Seifert, G. 3d Metal Nanowires and Clusters Inside Carbon Nanotubes: Structural, Electronic, and Magnetic Properties. *Phys. Rev.* B 2007, 75, 075410- 7.

[5] Somani, P. R.; Somani, S. P.; Umen, M. Application of Metal Nanoparticles Decorated Carbon Nanotubes in Photovoltaics. *App!. Phys. Lett.* 2008, 93, 033315-3.

[6] Lv, R.; Kang, F.; Gu, J.; Gui, X.; Wei, J.; Wang, K.; Wu, D. Carbon Nanotubes Filled with Ferromagnetic Alloy Nanowires: Lightweight and Wide-band Microwave Absorber. *App!. Phys. Lett.* 2008, 93, 223105-3.

[7] Kornev, K. G.; Halverson, D.; Korneva, G.; Gogotsi, Y.; Friedman, G. Magnetostatic Interactions Between Carbon Nanotubes Filled with Magnetic Nanoparticles *App!. Phys. Lett.,* 2008, 92, 233117-3.

[8] Zheng, H.; Wang, J.; Lofland, S. E.; Ma, Z.; Mohaddes-Ardabili, L.; Zhao, T.; SalamancaRiba, L.; Shinde, S. R.; Ogale, S. B.; Bai, F.; Viehland, D.; Jia, Y.; Schlom, D. G.; Wuttig, M.; Roytburd, A.; Ramesh, R. Multiferroic BaTiO3-CoFe2O4 Nanostructures. *Science* 2004, 303, 661–663.

[9] Yang, C. K.; Zhao, J.; Lu, J. P. Magnetism of Transition-metal/Carbon-nanotube Hybrid Structures. *Phys. Rev. Lett.* 2003, 90, 257203-4.

[10] Bin, Y.; Chen, Q.; Tashiro, K.; Matsuo, M. Electrical and Mechanical Properties of Iodine-Doped Highly Elongated Ultrahigh Molecular Weight Polyethylene Films with Multiwalled Carbon Nanotubes. *Phys. Rev.* B 2008, 77, 035419-7.

[11] Jeong, B. W.; Lim, J. K.; Sinnott S. B. Tensile Mechnical behavior of Hollow and Filled Carbon Nanotubes Under Tension or Combined Tension-torsion. *App!. Phys. Lett.* 2007, 90, 023102-3.

[12] Wang, L.; Zhang, H. W.; Zhang, Z. Q.; Zheng, Y. G.; Wang, J. B. Buckling Behaviors of Single-walled Carbon Nanotubes Filled with Metal Atoms. *App!. Phys. Lett.* 2007, 91, 051122-3.

[13] Ni, B.; Sinnott, S. B.; Mikulski, P. T.; Harrison, J. A. Compression of Carbon Nanotubes Filled with C60, CH4, or Ne: Predictions from Molecular Dynamics Simulations. *Phys. Rev. Lett.* 2002, 88, 205505-4.

[14] Karmakar, S.; Sharma, S. M.; Teredesai, P. V.; Sood, A. K. Pressure-induced Phase Transitions in Iron-filled Carbon Nanotubes: X-ray Diffraction Studies. *Phys. Rev.* B 2004, 69, 165414-5.

[15] Ye, Q.; Cassell, A. M.; Liu, H.; Chao, K. J.; Han, J.; Meyyappan, M. Large-scale Fabrication of Carbon Nanotube Probe Tips for Atomic Force Microscopy Critical Dimension Imaging Applications. *Nano Letters* 2004, 4, 1301–1308.

[16] Regan, B. C.; Aloni, S.; Ritchie, R. O.; Dahmen, U.; Zettl, A. Carbon Nanotubes as Nanoscale Mass Conveyors. *Nature* 2004, 428, 924–927.

[17] Ghosh, S.; Sood, A. K.; Kumar, N. Carbon Nanotube Flow Sensors. *Science* 2003, 299, 1042–1044.

[18] García-Vidal, F. J.; Pitarke, J. M.; Pendry, J. B. Silver-filled Carbon Nanotubes Used as Spectroscopic Enhancers. *Phys. Rev.* B 1998, 58, 6783–6786.

[19] Schlapbach, L.; Züttel, A. Hydrogen-storage Materials for Mobile Applications. *Nature* 2001, 414, 353–358.

[20] Wu, Z.; Chen, Z.; Du, X.; Logan, J. M.; Sippel, J.; Nikolou, M.; Kamaras, K.; Reynolds, J. R.; Tanner, D. B.; Hebard, A. F.; Rinzler, A. G. Transparent, Conductive Carbon Nanotube Films. *Science* 2004, 305, 1273–1276.

[21] Yoshida, N.; Arie, T.; Akita, S.; Nakayama, Y. Improvement of MFM Tips Using Fe-AlloyCapped Carbon Nanotubes. *Physica B-Condens. Matter* 2002, 323, 149–150.

[22] Alexiou, C.; Jurgons, R.; Seliger, C.; Iro, H. Medical Applications of Magnetic Nanoparticles. *J. of Nanoscience and Nanotech.* 2006, 6 (9/10), 2762–2768.

[23] Hummer, G.; Rasaiah, J. C.; Noworyta, J. P. Water Conduction through the Hydrophobic Channel of a Carbon Nanotube. *Nature* 2001, 414, 188–190.

[24] Margine, E. R.; Lammert, P. E.; Crespi, V. H. Reciprocal Space Constraints Create Real-space Anomalies in Doped Carbon Nanotubes. *Phys. Rev. Lett.* 2007, 99, 196803-4.

[25] Seifu, D.; Karna S. Nanomagnetics; ARL-TR-3790: U.S. Army Research Laboratory: Aberdeen Proving Ground, MD, April 2006.

[26] Seifu, D.; Hijji, Y.; Hirsch, G.; Karna, S. Chemical Method of Filling Carbon Nanotubes with Magnetic Material. *Journal of Magnetism and Magnetic Materials* 2008, 320 (3–4), 312–315.

[27] Prados, C.; Crespo, P.; González, J. M.; Hernando, A.; Marco, J. F.; Gancedo, R.; Grobert, N.; Terrones, M.; Walton, R. M.; Kroto, H. W. Hysteresis Shift in Fe-filled Carbon Nanotubes due to y-Fe. *Phys. Rev.* B 2002, 65, 113405-4.

[28] Wei, B. Q.; Vajtai, R.; Jung, Y.; Ward, J.; Zhang, R.; Ramanath, G.; Ajayan, P. M. Assembly of Highly Organized Carbon Nanotube Architectures by Chemical Vapor Deposition. *Chem. Mater.* 2003, 15, 1598–1606.

[29] Lowndes, D. H.; Geohegan, D. B.; Puretzky, A. A.; Norton, D. P.; Rouleau, C. M. Synthesis of Novel Thin-film Materials by Pulsed Laser Deposition. *Science* 1996, 273, 898–903.

[30] Pitt, C. W. Thin Film Technology. *Nature* 1977, 270, 16.

[31] Behrisch, R. *Sputtering by Particle bombardment* (ed.); Springer, Berlin, 1981.

[32] Keller, N.; Pham-Huu, C.; Estournes, C.; Greneche, J.; Ehret, G.; Ledoux, M. Carbon Nanotubes as a Template for Mild Synthesis of Magnetic CoFe2O4 Nanowires. *Carbon* 2004, 42, 1395–1399.

[33] Bozorth, R. M.; Tilden, E. F.; Williams, A. J. Anisotropy and Magnetostriction of Some Ferrite. *Phys. Rev.* 1955, 99, 1788–1798.

[34] 34 Benjamin, J. S. Mechanical Alloying. *Sci. Am.* 1976, 234, 40.

[35] Smith, D. L. Thin Film Deposition – Principle and Practice; McGraw-Hill, 1995.

[36] Chikazumi, S. Physics of Ferromagnetism, 2nd ed., Clarendon Press., Oxford, 1977.

[37] Stoner, E. C.; Wolfrath, E. P. A Mechanism of Magnetic Hysteresis in Heterogeneous Alloys. *Philos. Trans. R. Soc.* London 1948, Ser. A240, 599.

In: Carbon Nanotubes
Editor: Percy Szalkowski

ISBN: 978-1-62618-493-0

Chapter 5

On the Performance of Carbon Nanotubes in Extreme Conditions and in the Presence of Microwaves*

Julia B. Doggett and Ryan C. Toonen

Abstract

Using van der Pauw and microwave surface resi stance measurements, a seri es of temperature-dependent data sets from carbon nanotube (C NT) thin fi l ms have been measured. The test structures were fabricated using photolithography, E -beam evaporation, and a novel CNT network deposition technique. The sheet resistance and resistivity of each sample were recorded at temperatures ranging from 0– 60 °C with test currents ranging from 100 nA to 100 μA. These val ues demonstrated excellent l inearity, with no dependence on currents. At room temperature, the sheet resistance yielded a negative temperature coefficient (of approximately – 900 ppm/°C). Our objective is to analyze the effect that DC and microwave currents have on CNT thin films using a two-point measurement technique on Corbi no discs and in the presence of microwaves (8– 12 GHz) to determi ne at which temperatures a CNT thin film performs best and identify the ideal temperature range in which

* This is an edited, reformatted and augmented version of an Army Research Laboratory publication, ARL-M R-0836, dated January 2013.

a CNT microchip yields the maxi mi zed sheet resi stance. The results counter the trend found by the four-point measurements; a positive temperature coefficient was observed. This observation indicates the temperature coefficient of the material is actually dependent on the size of the sample tested.

1. INTRODUCTION

There is currently a strong military interest in the use of transparent conductors for communication applications. Materials that have been considered for use as transparent conductors include: transparent conducting oxides (TCOs), intrinsically conducting polymers (ICPs), graphene, elemental metal nanowires, and carbon nanotubes (CNTs). All of these materials offer advantages and disadvantages. TCOs offer excellent conduction and optical transmission characteristics but tend be very brittle and cannot be fabricated on flexible substrates. ICPs have good conduction and optical transmission properties, but are extremely sensitive to environmental conditions (such as temperature and humidity).

Graphene has recently attracted much attention within the research community because of its exotic physical properties. While it is very compatible with flexible substrates, it is very challenging to manufacture and its electrical properties for communication applications are still inadequate at this point in time.

Elemental metal nanowires offer excellent electrical properties are also very compatible with flexible substrates; however, they easily oxidize and require passivation coatings. In this study, we have tested the properties of CNT thin films [1– 3] for the purpose of evaluating their performance in communication electronics applications.

2. FOUR-POINT DC MEASUREMENTS

2.1. Setup

Four-point test structures [4, 5] were fabricated on top of sapphire substrates using photolithography [6] and electron-beam evaporation processes. A layer of chrome was applied as an adhesive between the substrate and the metal conduction layer.

Copper was applied to the chrome layer to reduce the resistance of the film. A gold cap layer was required to protect the copper layer from becoming oxidized and damaged. A thin film composed of a CNT network was deposited by an independent company called Nano-C© and returned almost test-ready. When returned, the wafers were diced by hand using a diamond scribe. Because two of the contacts were damaged during the dicing procedure, silver paint was applied to the sample to serve as improvised contact/probe-landing points.

Figure 1 shows the CNT thin-film structure.

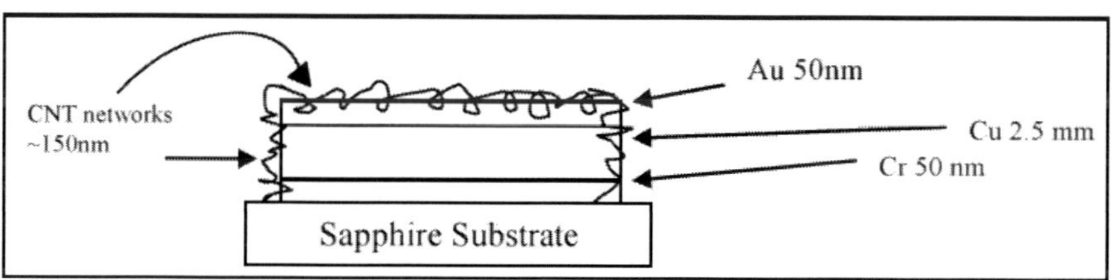

Figure 1. CNT thin film.

Nine temperatures were tested i n this experiment. Due to the seemingly strange behavior of the first set, the focus of the second set was shifted to the range of 0–10 °C (typical freezing range). The fi nal set was defined by the temperatures 0, 1, 2, 3, 6, 10, 20, 30, and 60 °C (commercial product test ceiling).

Test currents were sourced from 100 nA to 100 μA. Room temperature was roughly 24 °C and relative humidity was at 65%. The results were as followed.

2.2. Results

The calculated temperature coefficient for each temperature was negative, meaning that as the sample temperature increases, the sheet resistance decreases. This trend is shown in figure 2.

Figure 3 and table 1 show the temperature-dependent data of sheet resistance measured using the four-point technique. In figure 3, the sheet resistance was measured using test currents ranging in value from 100 nA to 100 μA.

These data show exceptional temperature stability. There is little self-heating present; the test current does not greatly influence the measured sheet resistance.

Table 1 lists the sheet resistance data points, measured using the limits of the test current range (100 nA and 100 μA), as a function of temperature, which ranged from 0 to 60^0C.

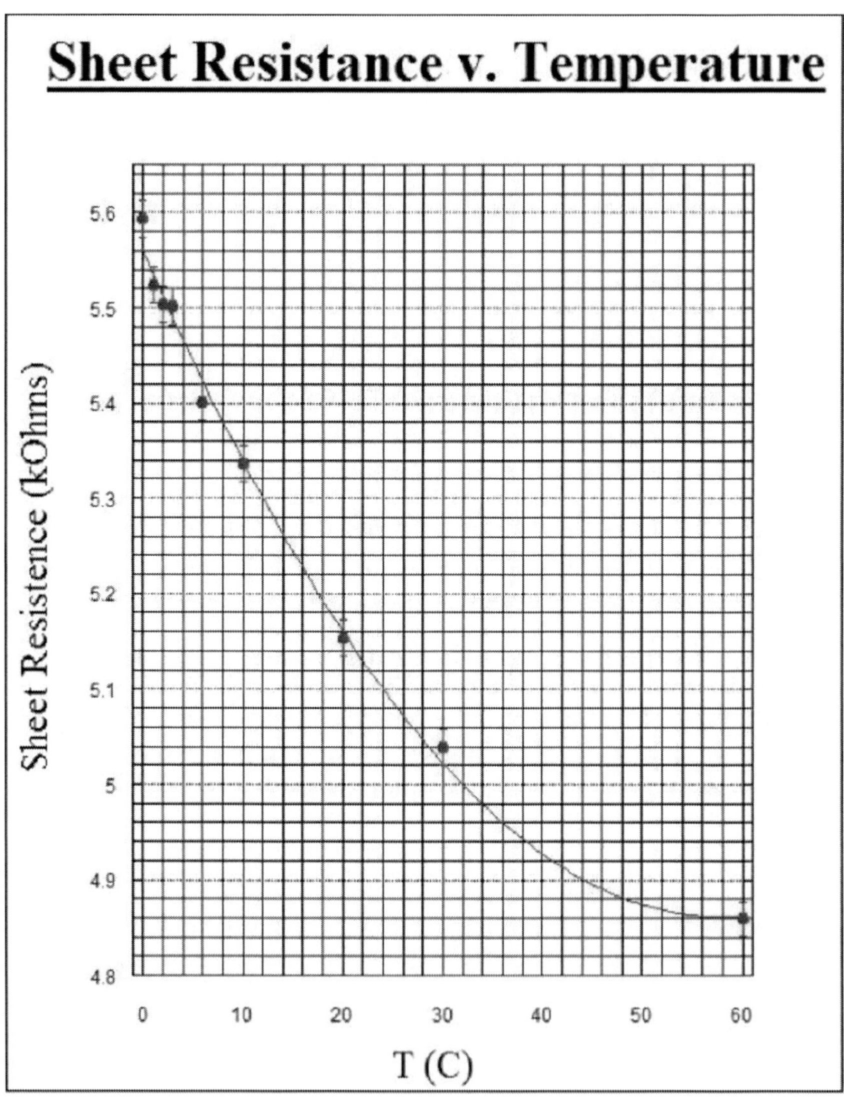

Figure 2. Overall decrease in sheet resistance as temperature increases.

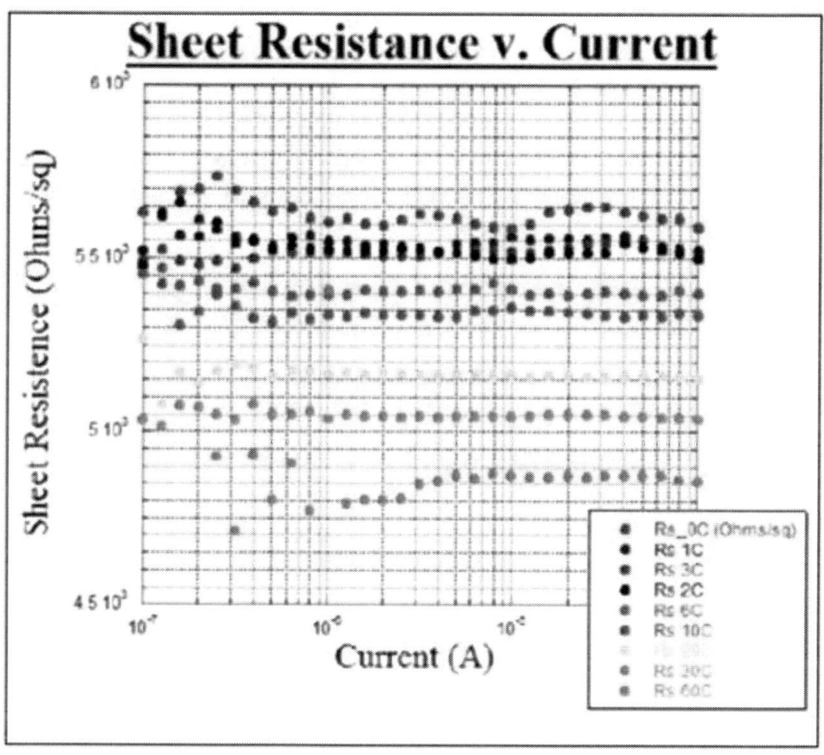

Figure 3. Temperature dependence of DC sheet resistance measured with test currents ranging from 100 nA to 100 μA.

Table 1. Temperature dependence of DC sheet resistance measured with test currents of 100 nA and 100 μA

T(C)	Input	Output (R_s) in ohms/sq	Input	Output (R_s) in ohms/sq
0 degrees	100 nA	5943.33	100 μA	5745.65
1 degree	100 nA	5523.30	100 μA	5529.14
2 degrees	100 nA	5478.57	100 μA	5502.93
3 degrees	100 nA	5490.78	100 μA	5501.04
6 degrees	100 nA	5455.97	100 μA	5400.80
10 degrees	100 nA	5267.03	100 μA	5336.04
20 degrees	100 nA	5265.23	100 μA	5153.57
30 degrees	100 nA	5034.37	100 μA	5038.89
60 degrees	100 nA	3295.08	100 μA	4859.40

These data were collected on the program LabView v.8.5 (using a code which performs a van der Pauw algorithm) and formatted into graphs on the program KaleidaGraph 4.1 (table 2).

The temperature coefficient represents the deviation of the sheet resistance due to a change in temperature. This value is calculated by the following equation but then converted into percentages for the purposes of this study:

$$T_c = (R_s/R_s(T)) * 100\%. \qquad (1)$$

Table 2. Temperature coefficient, from four-point measurement versus temperature

	Temperature Coefficient
at 20 °C	~0.09%
at 30 °C	~0.1%
at 60 °C	~0.15%

3. Corbino Disc Microwave Measurements

Corbino discs are a supplemental test structure used for sheet resistance characterization. A single Corbino disc is composed of an inner conducting circular contact and an outer conducting ring contact. In this study, the space between the two was occupied by a CNT mesh network.

An unexpected occurrence was observed when the testing was first initiated. Due to the high temperatures (approaching 310 °C) of the data set, the titanium-nitride cap began to melt with the other layers and started to smear. It was decided that the sample integrity was comprised, and these resistance measurements were regarded as invalid.

As a replacement film, a gold cap was used for testing, rather than titanium nitride, and the CNTs were installed underneath the metallic layers. Room temperature was 21.1 °C, and the relative humidity was at 65%. The test temperatures ranged from 296 to 400 K.

For the DC sheet resistance measurement, the voltages were sourced from 0.4 V to –0.4 V and from – 0.4 V back to 0.4 V. The reason for this selection in voltages is to determine if there were any detectable effects of hysteresis on the sweep downward. A DC current-versus-voltage measurement (taken at room temperature) of a Corbino disc with a diameter of 300 μm and a spacing of 45 μm is shown in figure 4. Resistance data were extracted from the slope of this curve, and the sheet resistance information was calculated by taking into consideration the geometric dimensions. Figure 5 shows a plot of sheet resistance values for Corbino discs of different diameters (d) and spacings (g). These measurements seemed to converge to approximately 800 ohms per square as a function of spacing.

The purpose of the microwave measurement was to gather further insight about how the thin films perform in a variety of communication scenarios. Similar temperatures were used when testing with microwaves (296 to 400 K). Microwaves ranged from 8 to 12 GHz (X band) and the signal power level was held constant at 1 mW. Figure 6 demonstrates the dependence of sheet resistance on microwave frequency. The temperature dependence of both the DC and microwave sheet resistance is shown in figure 7.

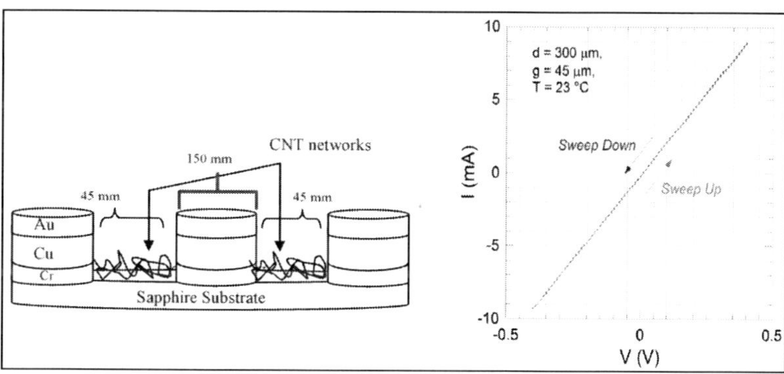

Figure 4. The graph shows there is no detectable effect of hysteresis on the downwards sweep and the contacts experienced very ohmic behavior.

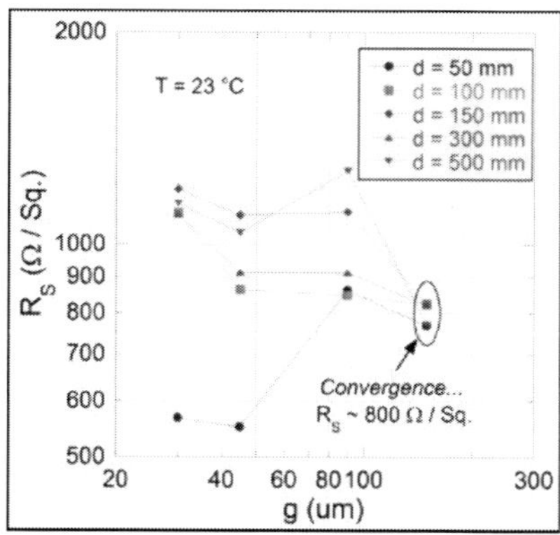

Figure 5. DC sheet resistance versus Corbino disc gap. The convergence of the different diameters at the sheet resistance of ~800 ohms/sq.

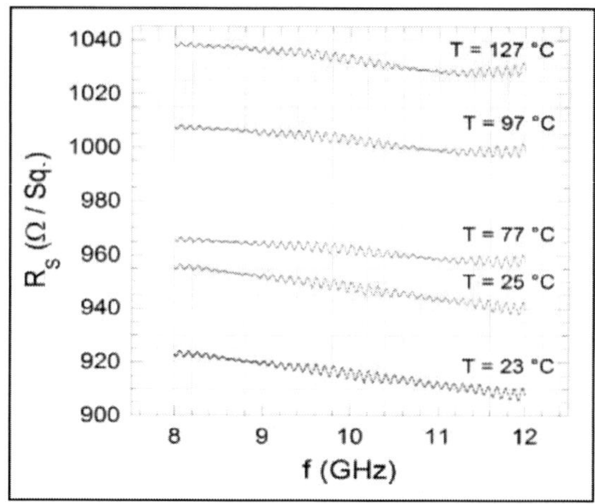

Figure 6. Microwave sheet resistance versus frequency.

From the cross-correlation values, we can assume there is some noticeable relationship between the DC and microwave (8 and 12 GHz) sheet resistance functions. The relationship is not absolute, but is noteworthy. This relationship reveals the amount of effect the microwaves have on the current (figure 7). This coefficient was calculated by the following equation [7]:

$$(f \star g)[n] \stackrel{\text{def}}{=} \sum_{m=-\infty}^{\infty} f^*[m]\ g[n+m]. \tag{2}$$

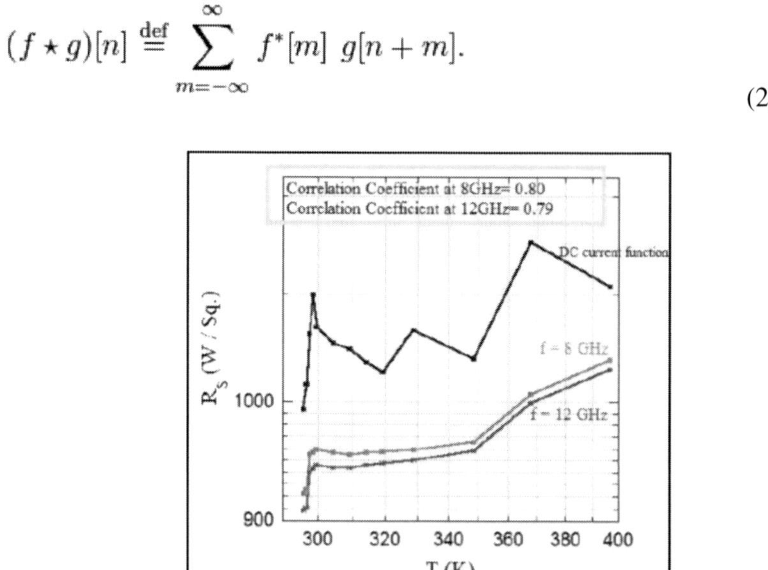

Figure 7. DC and microwave sheet resistance versus temperature.

Conclusion

In summary, the CNT thin films performed well in all of the conditions of this experiment. In the four-point measurement, there was some temperature-dependence when measuring sheet resistance, but the sample remained linear, displaying its stability and versatility in the presence of a range of currents. Although the Corbino disc two-point measurements were unpredictable, and at times incoherent, from those measurements, we were able to construct better contacts with the first two failures in mind. The obscured sheet resistances resulting from the first few sets of measurements could have been due to a native copper oxide that accumulated under the cap layer resulting in poor contact connection. The two-point tests actually yielded a positive temperature coefficient compared to the negative temperature coefficient of the four-point measurements. Most likely this discrepancy is a size effect resulting from heat-induced expansion of the CNTs. An additional microwave test was performed using 1 to 8 GHz (the L, S, and C bands) and similar behavior was observed, these data, however, were been omitted from this report due to time restraints. But the method of probing on-chip Corbino disc test structures placed in direct contact with pre-deposited CNT networks is an entirely new technique that can be credited to this study.

This body of work resulted from an eight-week summer internship sponsored by the George Washington University/Department of Defense (DoD) Science and Engineering Apprenticeship Program (SEAP).

References

[1] Nanocyl Web site. “Carbon Nanotubes,” 2009. http://www.nanocyl.com/en/CNT-ExpertiseCentre/Carbon-Nanotubes (accessed 11 July 2012).

[2] Jorgensen, T.; Grove-Rasmussen, K. Electrical Properties of Carbon Nanotubes. Scribd.com, 28 August 2000. http://www.scribd.com/doc/55420247/Electrical-properties-ofCarbon-Nanotubes (accessed 11 July 2012).

[3] Adams, T. A. II. Physical Properties of Carbon Nanotubes. 26 April 2000. http://www.pa.msu.edu/cmp/csc/ntproperties/ (accessed 11 July 2012).

[4] National Institute of Standards & Technology (NIST) Web site. "II. The Hall Effect," 15 April 2010. http://www.nist.gov/pml/div683/hall effect.cfm (accessed 2 July 2012).
[5] NIST Web site. "III. Resistivity and Hall Measurements," 15 April 2010. http://www.nist.gov/pml/div683/hall resistivity.cfm (accessed 27 June 2012).
[6] The Free Dictionary. "Photolithography." Farleç 2012.
[7] http://www.thefreedictionary.com/photolithography (accessed 11 July 2012).
[8] Wikipedia. "Cross-correlation." Wikimedia Foundation, 8 January 2012. http://en.wikipedia.org/wiki/Cross-correlation (accessed 9 August 2012).

INDEX

D

E

F

G

T

U

V

W

X

Y